**BRICKWORK
REPAIR AND
RESTORATION**

A fine example of nineteenth century gault clay brickwork in need of cleaning and repair.

Letheringsett · Norfolk

BRICKWORK REPAIR AND RESTORATION

W G Nash
formerly Head of Department of Construction Southampton Technical College
author of Brickwork Vols 1,2 & 3. Stanley Thornes

ATTIC
BOOKS

Attic Books
The Folly, Rhosgoch, Painscastle,
Builth Wells, Powys LD2 3JY

First published 1986
Second Edition 1989, Reprinted 1992

Copyright © W G Nash 1986 and 1989
© Frontispiece J Bowyer 1986 and 1989

British Library Cataloguing in Publication Data

Nash, W. G. (William George)
Brickwork repair and restoration.–2nd ed.
1. Buildings. Brickwork. Maintenance & repair.–
Manuals
I. Title
693'.21

ISBN 0-948083-09-3

All rights reserved. No part of this publication may
be reproduced, stored in a retrieval system or
transmitted in any form or by any means
electronic, mechanical, photocopying, recording or
otherwise, without the prior permission in writing
of the publisher.
No part of this publication may be copied or
reproduced by any means under licence issued by
The Copyright Licensing Agency Ltd.
This book is sold subject to The Standard Conditions of
Sale of Net Books and may not be sold in the UK below
the net price given by the Publishers on the book cover or
in their current price list.

Cover design by Jack Bowyer

Printed by J. W. Arrowsmith Ltd., Bristol

Contents

Author's preface and acknowledgements

1 Bricks and mortar
Materials
- Clay bricks — 1
- Sand lime bricks — 1
- Special purpose bricks — 2
- Site cutting of bricks — 5
- Fine aggregate or sand — 5
- Cement — 5
- Lime — 5
- Mortars — 6
- Recommended mortar mixes — 6
- Preparing the mortar mix — 6

2 Bonds and bonding
- Terminology — 7
- Bonding — 7
- Bonds — 9
- Cavity walling — 12
- Pitfalls in bonding — 12
- Brick backings — 13
- Decorative work — 16
- Matching brickwork — 19

3 Jointing and pointing
- Jointing — 20
- Pointing — 20
- Mortars for pointing — 21
- Types of pointing — 21
- Additives — 24

4 Cleaning brickwork
- Removing plant growth — 25
- Rendered walling — 25
- Assessing work to be done — 25
- Recognising defects — 26
- General — 31

5 Repairing brick facework
Remedial work — 32
- Bed joints — 32
- Sulphate attack — 32
- Frost action — 32
- Chimneys — 34
- Cracking in the walling — 34
- Corrosion of iron and steel — 35
- Unstable cavity walling — 36
- Damaged corbels etc. — 37
- Deterioration in parapet walls — 37
- Dampness in buildings — 38
- Bulging walls — 39
- Flood damage — 41
- Subsidence — 41
- Fire damage — 42

Summary — 42

6 Repairing and restoring brick features
- Fine axed work — 43
- Preparing voussoirs for an arch — 46
- Gauged and rubbed work — 48
- Camber arch in gauged work — 48
- Niches — 51
- Brick dressing to quoins and reveals — 55
- Tumbling-in courses — 55

7 Foundations and underpinning
- Subsidence — 56
- Remedial work — 57
  - Underpinning — 57
  - Piling — 59
- Structural repairs — 59

Glossary of terms — 60

Student exercises — 64

Preface

It would be true to say that among the population in general there is a marked preference for spending large amounts of money on leisure and entertainment, and a reluctance generally to allocate equivalent amounts to essential services such as housing. Such preferences have led to the virtual elimination of picturesque and architectural features in modern housing.

The functional properties of the latter have their supporters, but there is nevertheless a desire still in the majority of people to view and visit buildings of past eras and to admire the superb craftsmanship and skills evident from those bygone days. The huge numbers of visitors to our stately homes bear witness to this fact. One cannot help but wonder what future generations will inherit from us in craftsmanship terms.

Fortunately, however, there are still far-sighted people around who are to some degree safeguarding our heritage. This does mean, of course, that these much-visited (and other) buildings have to be carefully looked after, and it would be foolish not to acknowledge that the costs of this increase year by year.

Such magnificent buildings must be looked upon in the same way that doctors look upon their patients. When they are young they are robust and will resist most attacks but as they mature they need looking after, and when they grow old they need nursing and, occasionally, intensive care. This attitude must be maintained when carrying out repairs and restoration to any walling in old buildings, and careful planning is needed before any major decisions are made in relation to renovation.

The major problems that should be appreciated lie in the decreasing availability of some materials and also the scarcity of skilled craftsmen to carry out any repairs. This class of work is not for the mediocre operative.

Fortunately, there are still building companies around who are prepared to budget for adequate training of apprentices, so perhaps we should not be too pessimistic. Nevertheless we must hope that economics do not overtake us to the extent that such companies fall by the wayside and make it even more difficult to maintain the magnificent buildings which exist throughout the country.

Acknowledgement

My special thanks are due to Mr. Richard Brazier, a principal member of Brazier and Sons Ltd, Millbrook Road, Southampton, for his assistance in the preparation of this book. His company has made a valuable contribution to the repair and restoration of old buildings in the South of England.

W. G. Nash
Southampton 1984

1 Bricks and mortar

Materials

Clay bricks

Bricks may be classified as being among the most durable of building materials. Well burnt facing bricks will withstand the ravages of the weather extremely well and in spite of being subjected to rain, frost, heat and so on will retain their colour, texture and strength in a most remarkable way. There are numerous examples of brickwork throughout the country which have survived for centuries and are testimonies to the durability of bricks.

A good quality brick is made from a suitable clay, well moulded, thoroughly dried, and then hard burnt in order to change the chemical structure of the clay into non-soluble and inert compounds.

All bricks were at one time hand-made and most areas of the country had their local brickworks. This gave a wide variety of standards and types ranging from the very hard and dense bricks, such as the Staffordshire Blues of the Midlands and Accrington Reds of the North, through the less dense bricks of the Midlands and South, to the extremely soft and sandy bricks, used for gauged work, of the Bracknell area. The colours are just as varied and range from the nearly black, blues, browns, reds, yellows and nearly white. The colours depend upon the chemical constituents of the clays. For example, those with a high iron content will turn blue and red; those with less iron, like the yellow clay or the malms found in the London and Kent areas, will produce the yellow stocks; and the gault clays, such as those which are found in Cambridge and the New Forest areas with no iron content, will produce the very light coloured, almost white bricks.

The general processes for hand-made bricks involve the digging of the clay, allowing it to weather before pugging it with water in a pugmill, and turning it into a pliable mixture which allows the clay to be readily moulded. The moulded or green bricks must then be thoroughly dried before being burnt.

The burning may be carried out in an intermittent kiln such as a clamp or Scotch kiln. In the clamp the green bricks are stacked in walls or bolts with spaces in between to receive the fuel. The outside of the clamp is covered with underburnt bricks from a previous burning to protect the bricks in the kiln. This method produced a good, hard burnt brick, rather irregular in shape with a pleasant mottled finish. Unfortunately this method is rapidly disappearing mainly because of the decline in skilled labour who are capable of setting a clamp; and also because of its uneconomically high labour content.

The Scotch kilns are very similar to the clamps but they have four walls in which the green bricks are stacked, access being made through an opening at one end of the kiln which is built up temporarily while the bricks are being burnt. There is much less wastage with this type of kiln and hard, well burnt bricks are produced. There are still many of these kilns in operation, but because of the high labour costs the products are usually more expensive than their machine-made counterparts.

There is, however, quite a lot of machinery used in the brick-making industry such as excavators for extracting the clay, the pugmills for pugging the clay, presses for providing the pressed facings and the wire cutting frames for cutting the clay after extrusion. These bricks have a rough texture finish and have no frogs. Often, however, after the green bricks have been cut with the wire frame they are then placed in a mould and pressed and sold as hand-made facings and are a first class product.

As the brick-making industry required such a high labour content it was evident that the supply would not meet the demand, and this, together with its high costs, led increasingly to mechanical production. It was necessary to find suitable clays in sufficiently large quantities to allow this to be done. Investigations found that the London Blue Clays, which contained a natural fuel in the form of lignite, lent itself to moulding under very high pressures. This meant that there was only a minimum of water content, therefore the drying could be done more rapidly. Large deposits of this clay were discovered in the Bedfordshire area and from these the Fletton bricks were produced in vast quantities.

This particular type of clay can be dug and transported (often by moving belts) to the grinding mills and from there directly to the presses where it is moulded under great pressure into the green bricks which are roughly 10 per cent larger than the finished product to allow for shrinkage. The green bricks are then burnt in a continuous kiln, called a Hoffman kiln, which consists of a number of chambers, usually sixteen, each with heavy dampers which allow the fire to be transferred from one chamber to the next. All the chambers are connected to a central flue. As the fire passes round the kiln so the chambers in front of the firing zone are loaded with the green bricks and the temperature increased steadily until the bricks are ready for burning. The chambers behind the firing zone are allowed to gradually cool down for the unloading of the burnt bricks. The clay contains most of the fuel necessary for the burning but a small amount of pulverised coal (about 50 kg per 1000 bricks) is added to boost the final stage of the burning process. In addition to completing the chemical change in the clays it gives the bricks their colour.

Sand lime bricks (calcium silicate bricks)

These are made from a mixture of lime and sand in the proportion of about 1:8 to which is added a very small quantity of water. This mixture is moulded into the required shapes under great pressure and loaded on to flat trucks which in turn are pushed into an autoclave. When the autoclave is fully loaded a huge steel cap is bolted on to the end and steam is then injected until a pressure of 110 kN/m^2 is reached. This is maintained for about 8 hours. The pressure combined with the high temperature fuses the bricks into excellent building units.

Sand lime bricks are divided into four classes:

1 Bricks for special purposes. These are produced for use in work where a high strength is required, or where walls are subjected to continuous water saturation or are liable to be exposed repeatedly to temperatures below freezing.
2 Bricks which are suitable for general facing work.
3 Bricks suitable for general facing work in mortars other than strong cement mortars.
4 Bricks which are only suitable for internal work in mortars other than strong cement mortars.

BS 187 specifies the minimum requirements for sand lime bricks including their dimensions, appearance, crushing strengths, transverse strengths, and drying shrinkage limits.

From the foregoing it is apparent that there is a wide variety of bricks available for use throughout the country. Nevertheless, it may be difficult to trace a particular type needed for specialist restoration work, and in such cases it is recommended that the following bodies be contacted for further information:
The Brick Development Association
The Sand Lime Brick Association
The Building Centre

Special purposes bricks

Most brick manufacturers carry quite a variety of standard special shaped bricks such as:

Squints (Figure 1)
Single and double cants (Figure 2)
Single and double bullnose (Figure 3)
Plinths (Figure 4)
Plinth internal and external returns (Figure 5)

In addition to these standard special brick shapes, most manufacturers will also be prepared to produce purpose-made units for special work, but it is important to bear in mind that the manufacturer will require time to prepare the moulds, mould the clay, dry and burn the bricks. He will have to be supplied with specific dimensions of the finished products so that he can allow for drying shrinkage in their making. These specially made products will be expensive items, the cost mainly depending upon the total numbers required. If a multi-brick unit is required to be specially made, for example a camber arch, then the arch should be drawn full size and the thickness of joints clearly shown so that the manufacturer can produce moulds for the arch bricks, which in the case of the camber arch will have to be individually moulded (see Chapter 6).

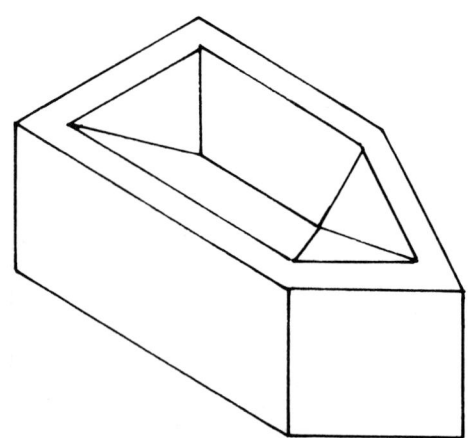

Figure 1 Squint

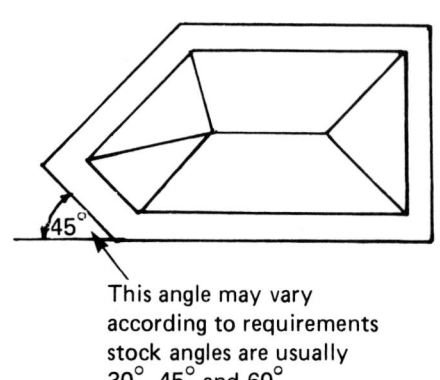

This angle may vary according to requirements stock angles are usually 30°, 45° and 60°

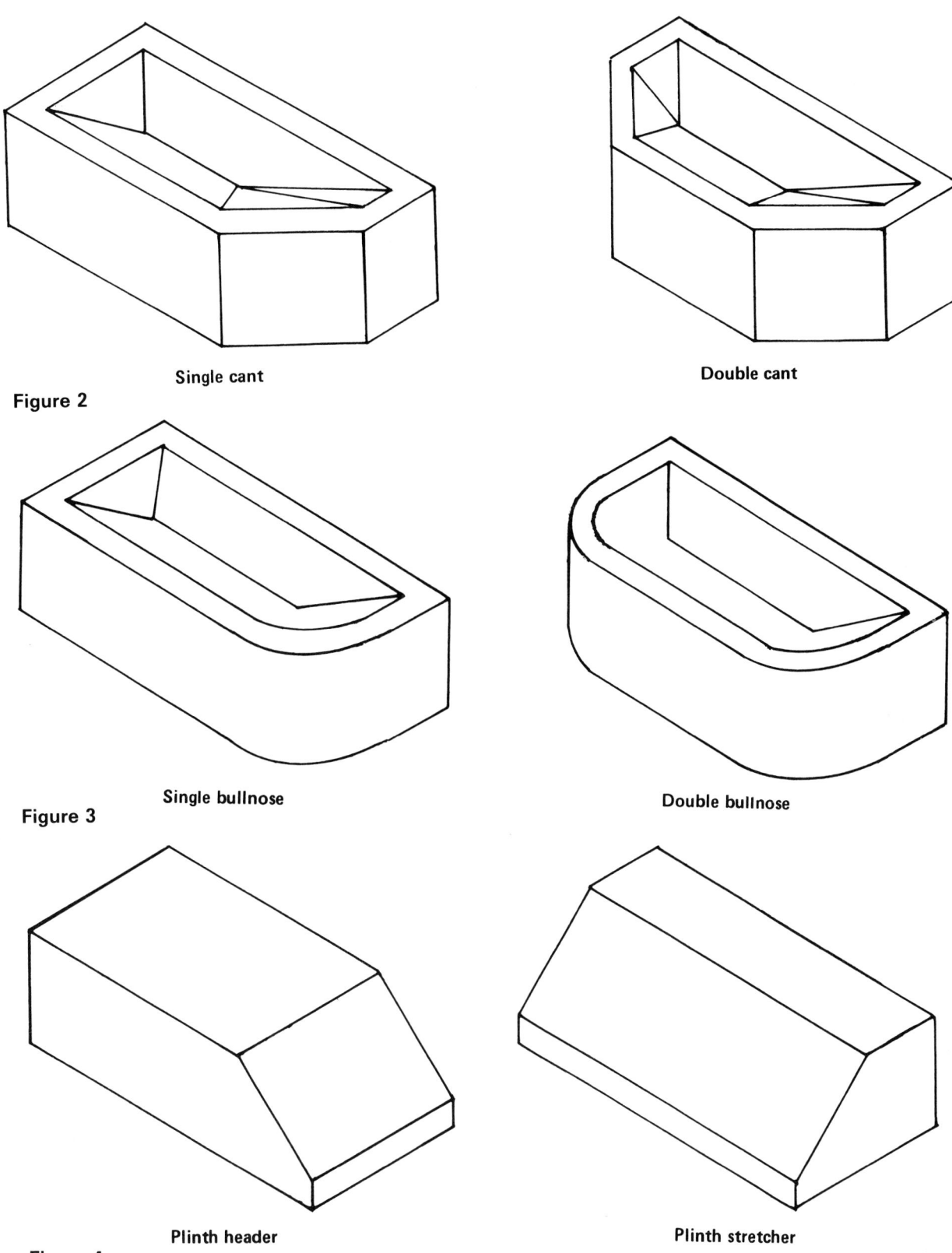

Figure 2 — Single cant / Double cant

Figure 3 — Single bullnose / Double bullnose

Figure 4 — Plinth header / Plinth stretcher

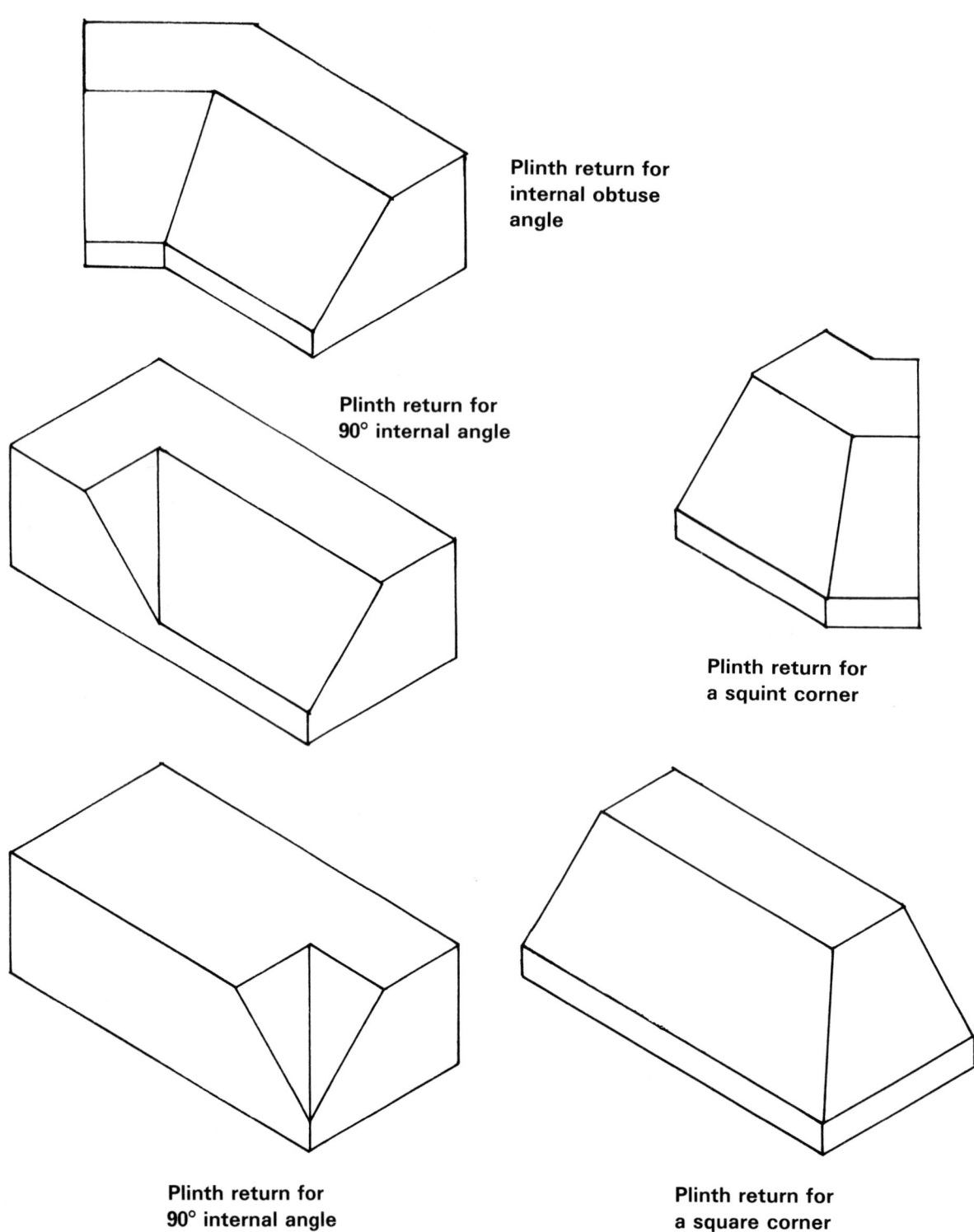

Figure 5

Site cutting of bricks

Most bricks may be cut by normal hand tools but usually this is often wasteful and costly in labour. Mechanical saws are readily available and most shapes may be obtained from these machines. The major problem with all cutting is that if the bricks to be cut are of a sand faced texture, and the face of the brick is to be cut (such as a squint or a cant brick), then the resulting surface left from the saw blade will be smooth and not have the same sandy surface as the remainder of the brick surface area. This factor has to be carefully considered when deciding whether to cut special bricks on site or to have them specially made by a brick manufacturer.

Fine aggregate or sand

The material which passes a 5 mm sieve is called sand and varies in content and colour quite considerably from area to area. For general brickwork it is quite usual for the sand to contain a high content of loam or clay (which are very fine particles) and this makes the mortar more plastic for working. While these fine particles will reduce the strength of the mortar this is not generally considered a major defect as high strengths are not usually necessary in mortars, and, in fact, may even be harmful (see the section on mortars).

Colours of sands vary and if a mortar is to be matched with existing work then samples should first be taken and test pieces built and allowed to dry and set before actual repairs or restoration commences.

These loamy sands may also contain organic impurities which can cause staining. Pyrites, for example, will give a rusty stain on the joints and may even penetrate into the bricks in severe cases.

A simple method of checking if a sample of sand contains some organic impurities is to put about 80 ml of sand in a measuring jar, or a bottle, pour in a 3 per cent solution of sodium hydroxide (caustic soda) up to the 120 ml mark, then shake thoroughly and leave for 4 hours. The colour of the liquid above the sand will indicate the possible presence of impurities. If it is a dark brown in colour then the sample is suspect; if it is a straw colour then it can be assumed to be free of impurities. This is not a quantitative test but purely a rule of thumb test to see if the sample is suspect or not.

For some work of a specialised nature a washed sand may be specified. This is a sand which has been washed through sieves thus removing all the fine particles of clay and loam.

Cement

Ordinary Portland cement or Portland Blast Furnace Cement are the types most commonly used for mortars and are made by burning a mixture of chalk (or limestone) and clay in a rotary kiln and then grinding the resultant clinker in a ball mill to a very fine powder.

BS 12 specifies the requirements for Portland cement and deals with the fineness, strength in compression, soundness, setting times and chemical composition.

Limes

The raw material from which building lime is obtained is either chalk or limestone. When this is burnt at a high temperature it is transformed into quicklime which, in this state, is unstable and unsuitable for building use. It is, therefore, stabilised by adding water to it. This process is called slaking and it raises the temperature of the quicklime and causes it to expand. The resultant material is hydrated lime and is ready for use in mortars or plasters. Limes set by a process of carbonation, in which the lime absorbs carbon dioxide slowly from the atmosphere.

Limestones vary in composition in different parts of the country, ranging in colour from white to grey and each producing a different type of building lime, which can be broadly placed into one of three categories:

1 *Non-hydraulic* (will not set under damp conditions)
2 *Semi-hydraulic* (will partially set under damp conditions)
3 *Emminently or magnesian limes* (which will set under damp conditions in the same way, though to a lesser extent, as Portland cement)

The pure limes may be mixed and allowed to stand until required for use, but the hydraulic limes should not be allowed to stand for more than 4 hours after mixing before using, otherwise they will lose their hydraulic properties.

Mortars

Mortars should have the following properties:

1 Be easily workable. This may readily be obtained by using lime. The higher the proportion of lime the better the working qualities.

It should also have good water retentivity particularly for bricks of high suction rate or if the mortar contains cement to ensure that the mortar stays wet long enough for the cement to develop its strength.

2 It should stiffen enough to avoid any delay in the building work. This can be controlled to a large extent by the craftsman by wetting the bricks, although the degree to which the bricks are wetted before use varies to suit the type of brick, the properties of the mortar and the weather.

3 It should set properly. The development of strength is a continuation of the stiffening process. Non-hydraulic lime and sand mixes have no real set but harden slowly by carbonation from the surface. If a cement is added to the mix it will increase the rate of hardening and will take place throughout the body of the joint and not just at the surface. The final strength needed in a mortar will vary according to the type of work. It is, however, interesting to note that the strength of brickwork does not always increase with the strength of the mortar.

4 It should be resistant to rain penetration which depends on getting tight joints and a good bond between the mortar and the bricks. Bricks with dusty surfaces that are not properly wetted will give a poor bond; as will the use of mortars which are unable to retain their water against the suction of the brick.

All mortars are subject to drying shrinkage which causes fine cracks to occur and which may allow penetration of rain in severe conditions. However, weaker mortars may give better results since the more porous mortars tend to absorb more of the water instead of letting it pass right through the wall.

5 It should be durable by being resistant to frost. While our winters are not cold by some standards we often have alternations of wet and frosty weather which puts a severe strain on building materials. All mortars are sensitive to frost during hardening but hardened mortars are only sensitive if weak.

They should also be durable by being resistant to sulphates. Some bricks contain enough soluble sulphates to cause trouble under damp conditions. The possible effects on bricks and on mortar are quite different. The effect on the bricks is an unsightly white efflorescence which appears soon after the building is completed and which should, if sufficient attention has been paid to such details as flashings and damp-proof courses, gradually disappear. In exceptional cases, however, the sulphates can cause the bricks to break up or flake.

With mortars, under persistently damp conditions, the sulphates from the bricks may cause a chemical reaction to take place between them and the Portland cement. This may be shown by a marked increase in volume which may cause the mortar to crumble or the brickwork to expand, or both.

Conditions favouring this reaction are to be found in parapet walls, copings or walls subject to severe exposure.

Recommended mortar mixes
A cement/aggregate mortar should be used only where a dense strong mortar is essential, for example, where engineering bricks are used for carrying heavy loads or in special cases for construction below ground damp-proof courses. For work of this type the workability of the mortar may be increased by adding lime up to one-fourth part of the volume of Portland cement. The usual ratio of the cement/sand mix would be 1:4.

Cement/lime/aggregate mortars

(a) A composition of 1:1:5 or 1:1:6 cement:lime:aggregate is recommended for use with all normal types of construction that are likely to be exposed to severe conditions. This type of mortar is not recommended for use with sand lime or concrete bricks that have a drying shrinkage greater than 0.025 per cent.
(b) Mortar composed of 1:2:8 or 1:2:9 cement:lime:aggregate is suitable for building normal brickwork except where the exposure conditions are likely to be severe. This mix is suitable for sand lime and concrete bricks.
(c) Mortar composed of 1:3:10 cement:lime:aggregate is recommended for internal work only.

Preparing the mortar mix
For cement mortars the cement and aggregate should be thoroughly mixed together in a dry state and then the water added until the right consistency is obtained.

For cement lime mortars, the lime and sand should be mixed together first and water added to give the right consistency. This mix may be kept for a long time provided care is taken to prevent it from drying out. This may be done by covering the heap with sacks which are kept damp or by any other suitable means. Immediately before use the cement is thoroughly worked in and water added to give the right consistency. This mortar should be used up within 2 hours otherwise the set of the cement will be partially destroyed.

2 Bonds and bonding

Terminology
The following are terms which are in common use and applied to walling in general:

Header   A brick laid with its narrow end to the face of the wall.
Stretcher   A brick laid with its length along the face of the wall.
Course   A complete layer of bricks.
Bed joints   The horizontal joints between the courses.
Perpends   The vertical joints between the bricks, also referred to as the cross joints.
Lap   The distance between perpends on consecutive courses.
Fair faced work   Walling which is to be left neat and clean.
Closer   A quarter bat or brick.
Half bat   A half brick.
Three-quarter brick   Three quarters of a brick.
Tie brick   The brick which ties in the internal angle at a corner, junction wall, buttress wall or pier and so on.
Quoin   An external corner of a building.
Stopped end   The end of a wall which is square and has no return wall at its extreme point.

Bonding
Bond patterns are essential for any wall which is intended to carry any loading as they ensure even distribution of the mass throughout the whole length of the walling thus avoiding, as far as possible, any structural failure due to the loading being concentrated at one point in the wall (Figure 6). In addition to distribution of the mass, stability is also maintained in the walling by correct bonding at quoins, attached piers, junction and separating walls thus ensuring that they are well tied together (Figures 7 and 8).

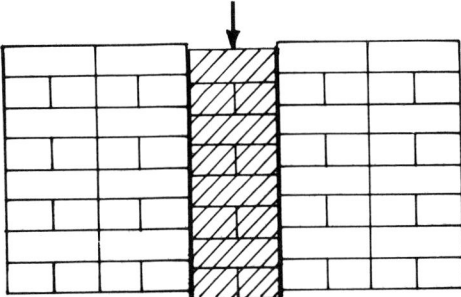

The shaded portion of the walling takes all the load and has a tendency towards more settlement than the rest of the walling

Uneven load distribution due to poor bonding

Figure 6

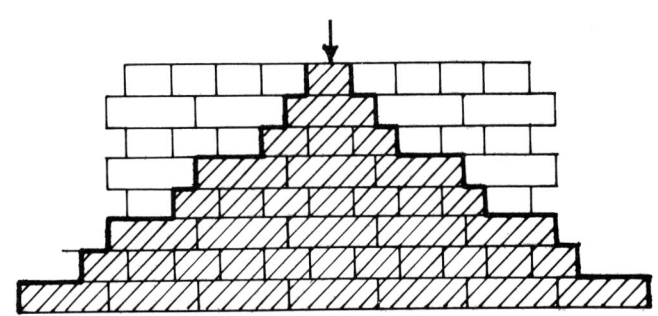

The shaded portion shows the loading being distributed over a large area of walling

Even load distribution due to good bonding

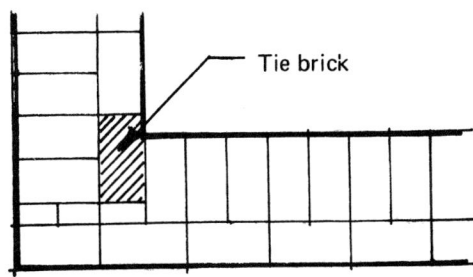
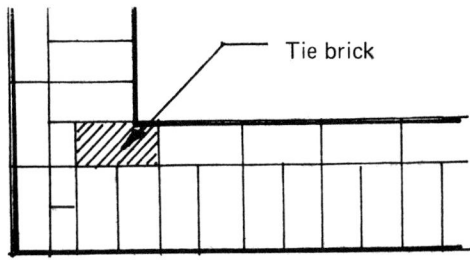

Plans showing the tie bricks bonding the walls together
Figure 7

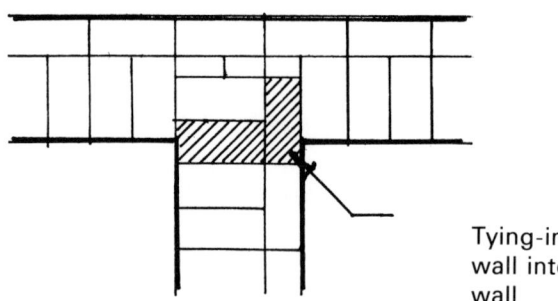

Tying-in a junction wall into a main wall

Figure 8

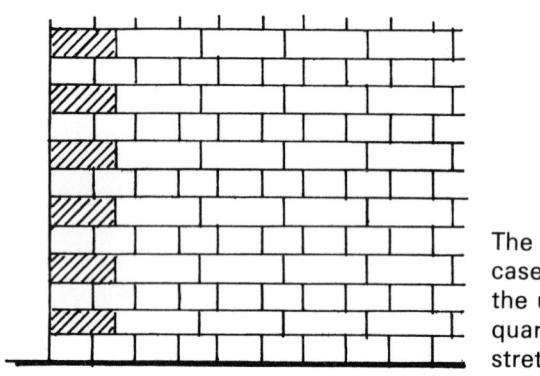

The bond in this case is formed by the use of three-quarter bats in the stretcher courses

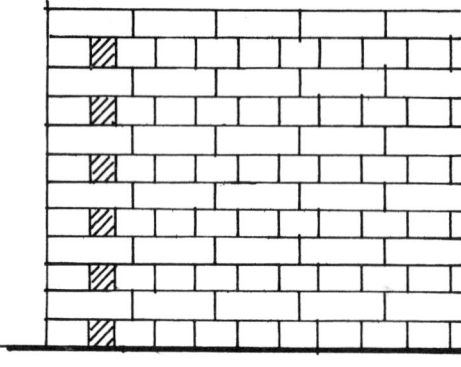

The bond formed by the introduction of closers in the header courses

Figure 9

While some bonds are reputed to be stronger than others there is, in actual fact, little difference between them. A well bonded wall will carry any normal loading quite safely.

The bonding of brickwork, however, is not wholly confined to strength requirements as very often a particular bond will be introduced for its appearance. In addition to these bonds, decorative patterns may be incorporated into face walling, and in this way plain walling may be given an interesting façade by the introduction of such architectural features.

In order to maintain the strength and appearance of walling certain principles should be applied so far as possible to bonding:

1. The correct lap should be set out and maintained by the introduction of a closer next to the quoin header, or alternatively starting each stretcher course with a three-quarter bat (Figure 9).
2. The perpends or cross joints in alternate courses should be kept vertical.
3. There should be no straight joints, i.e. no vertical joints should coincide in consecutive course or, where they are unavoidable, must be kept to an absolute minimum.
4. No closer should be built into a wall except next to a quoin header.
5. Junction walls and quoins must be securely bonded together with the aid of tie bricks.

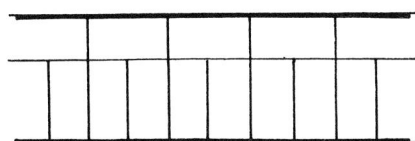 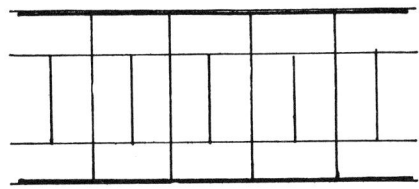

Plan showing the sectional bonding of walling

Figure 10

 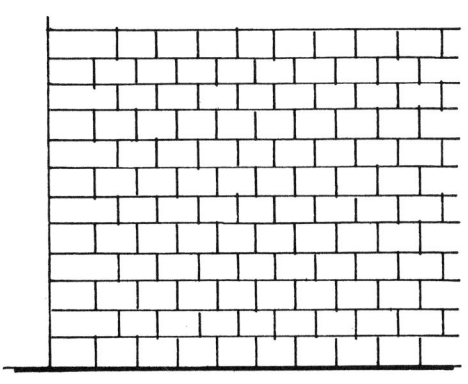

Stretcher bond

Figure 11

Header bond

Figure 12

6   The bricks in the interior of thick walls should be laid headerwise so far as possible.
7   Sectional bond should be maintained across the wall, i.e. the bond on the back of the wall should be in line with the bond on the face of the wall (Figure 10).

Bonds

It is possible to devise an infinite number of arrangements of bricks but there are certain bonds which are commonly used and easily recognised.

Stretcher bond

This bond is mainly used on walls of half brick in thickness. All the bricks are laid stretcherwise and the courses half bond with each other. The bond is formed by the introduction of a half bat, the stopped end on alternate courses or by the return header at a quoin (Figure 11).

Header bond

The bricks are all laid as headers and the bond being formed by the introduction of a three-quarter bat at the quoin or stopped end on alternate courses. This bond is used mainly on curved work, tall chimneys and footing courses (Figure 12).

English bond

This bond has alternate courses of headers and stretchers. The bond may be formed by either introducing a closer next to the quoin header or by the introduction of a three-

English bond

Figure 13

Flemish bond

Figure 14

Dutch bond

Figure 15

English garden wall bond

Figure 16

quarter bat at the end of each stretcher course (Figure 13).

Flemish bond
In this bond headers and stretchers are laid alternately in each course with the headers of one course laid centrally over the stretchers in the course below (Figure 14).

Dutch bond
This is a very popular bond and quite often mistaken for English bond because of its similarity in that it consists of alternate header and stretcher courses. However, in the Dutch bond no closers are used and the bond is formed by starting each stretcher course with a three-quarter bat. In addition to this the stretcher courses are laid half bond with each other, which is effected by introducing a header next to the three-quarter bat on alternate courses. The perpends in this bond follow each other diagonally across the wall in unbroken lines (Figure 15).

Garden wall bonds
These are decorative bonds which have been mainly derived from the English and Flemish bonds. Apart from their decorative feature they have another distinct advantage in that they can be used to build one brick thick walls with a fair face on both sides. Some types of bricks vary quite considerably in length and as the garden wall bonds require fewer headers it is possible to adjust the stretchers between the headers so that a fair face is achieved on both sides of the wall.

Flemish garden wall bond

Figure 17

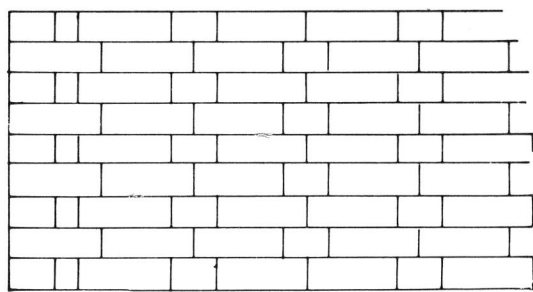

Monk bond

Figure 18

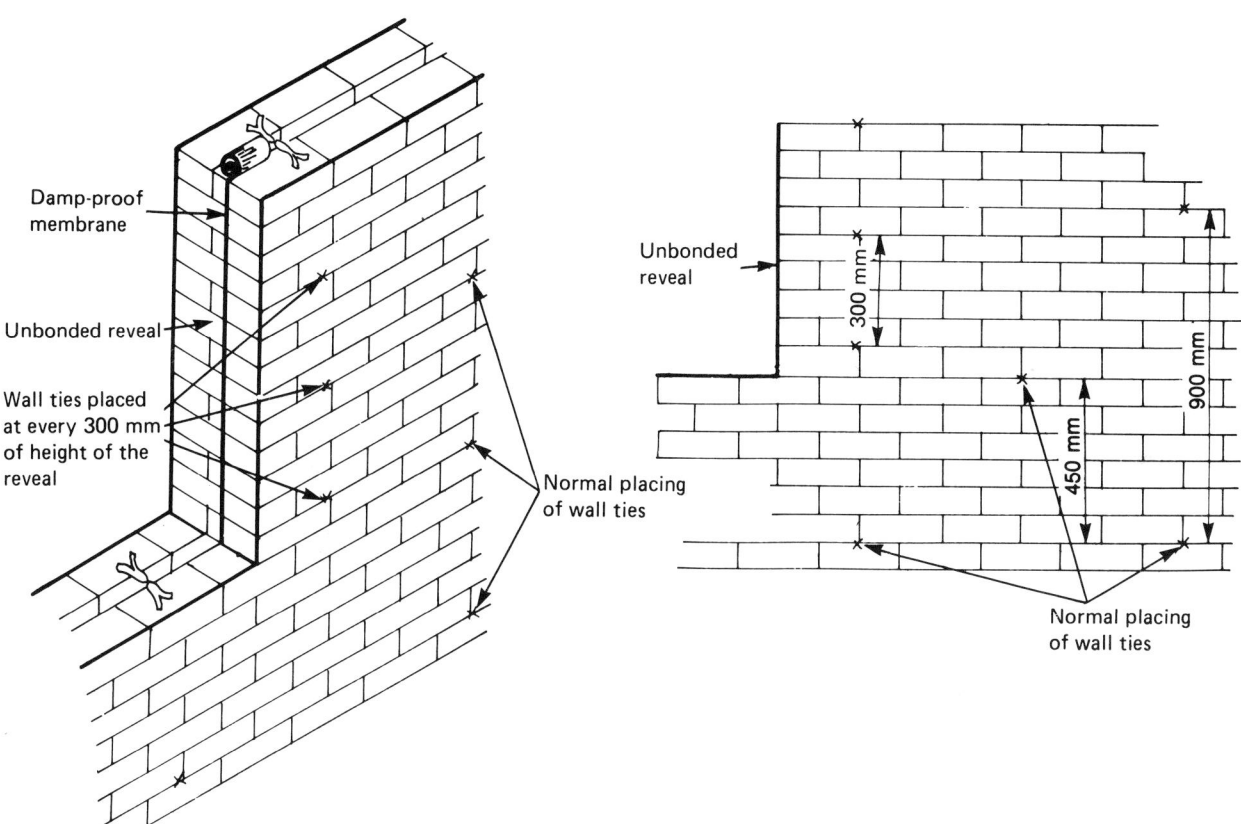

Placing wall ties at unbounded reveals

Figure 19

English garden wall bond
This bond consists of three and sometimes five stretcher courses to each header course (Figure 16).

Flemish garden wall bond
This bond has three and sometimes five stretchers to each header in every course, the header in one course being placed over the central stretcher in the course below (Figure 17).

Monk bonds
Monk bonds have two stretchers to each header in each course with the headers being placed over the joints between the stretchers. There are, however, several

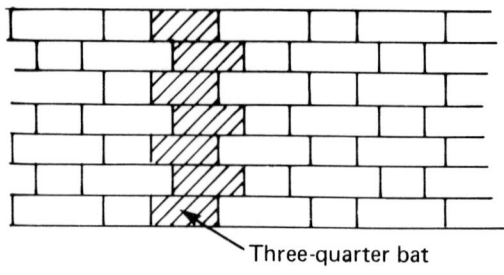

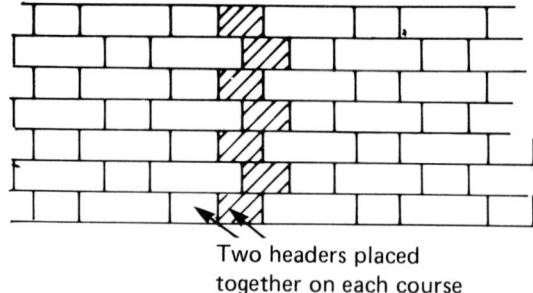

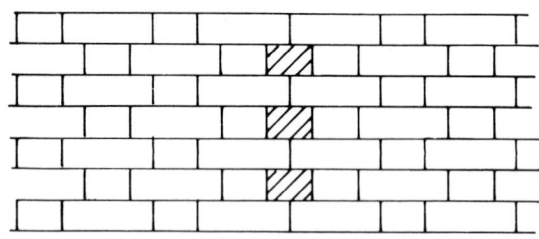

Typical broken bonds in Flemish bond

Figure 20

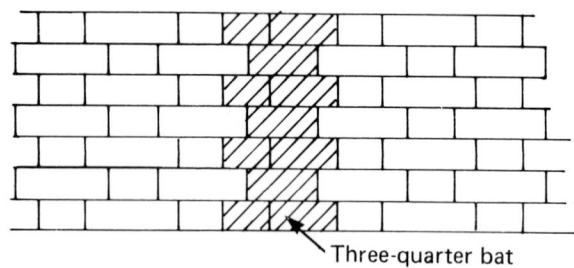

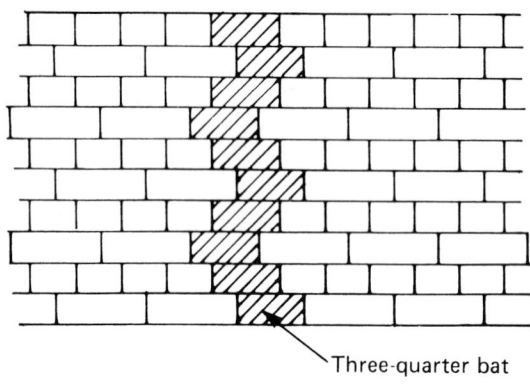

Typical broken bonds in Dutch bond

Figure 21

variations in the methods of starting the bond from the quoin or stopped end (Figure 18).

Cavity walling

This is mentioned here, not as a particular bond, but as walling, as so often cavity walling is assumed as being two half brick walls separated by a 50 mm cavity, but this is far from the truth. The inner leaf of the cavity wall may be of any thickness which is necessary to carry the live and dead loads of the building while the outer leaf may be a half brick wall separated from the inner leaf by a 50 mm cavity. The leaves should be stabilised by the introduction of wall ties which should be placed not more than 900 mm apart horizontally and 450 mm vertically.

They should be placed so that they are staggered over the walling. At unbonded reveals the vertical height of the wall ties should not exceed 300 mm (Figure 19).

Pitfalls in bonding

(a) When setting out bonds it is common practice to start from each end of the wall with the same brick, either a header or stretcher, and work towards the centre of the wall. Very often the length of the wall does not work out to complete brick size and a broken bond has to be introduced. Typical examples of these are shown in Figures 20 and 21.

(b) Where window openings are situated in walling the perpends should be set out at the base of the wall

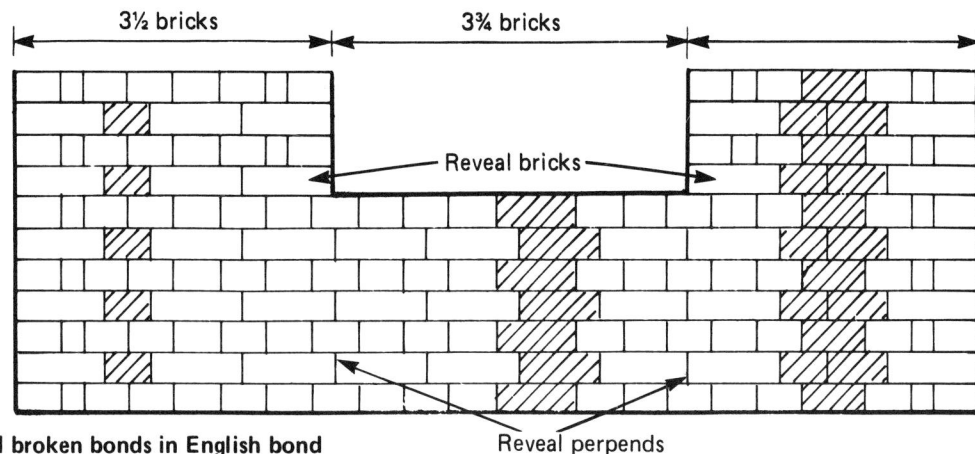

Three typical broken bonds in English bond

Figure 22

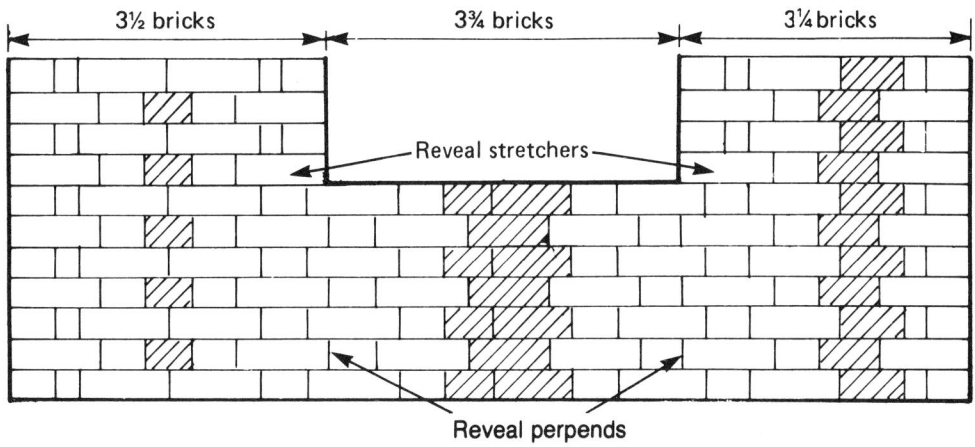

Three typical broken bonds in Flemish bond

Figure 23

(Figures 22 and 23) and this has to be done with care.

This is further complicated by the difficulty that is encountered at quoins where the bond on one face, for example, is a stretcher course and the same course on the return wall it is headers (Figure 24).

(c) In some cases it may be possible to introduce reverse bond in some piers rather than have a broken bond. Figure 25 shows examples of this bond, and it will be seen that this method of bonding is far more economical in materials and labour than introducing cut bricks in the middle of the pier, but, of course, the bond on opposite reveals will be different. This feature will be noticed where the brickwork above the opening is carried over by means of a steel lintel as one side of the opening will have a straight joint, which may not be desirable (Figure 26).

(d) When setting out bonds for walling which have plinth courses at their base the neat work should be set out above the plinth courses first and the bond for the plinth courses worked downwards (see Figures 27 and 28).

Brick backings

Composite walls may be built of a combination of materials; the external faces being constructed with a traditional brick or stone to blend in with the local surroundings, while the inside of the wall may be of lightweight bricks or blocks. Such walls may also be

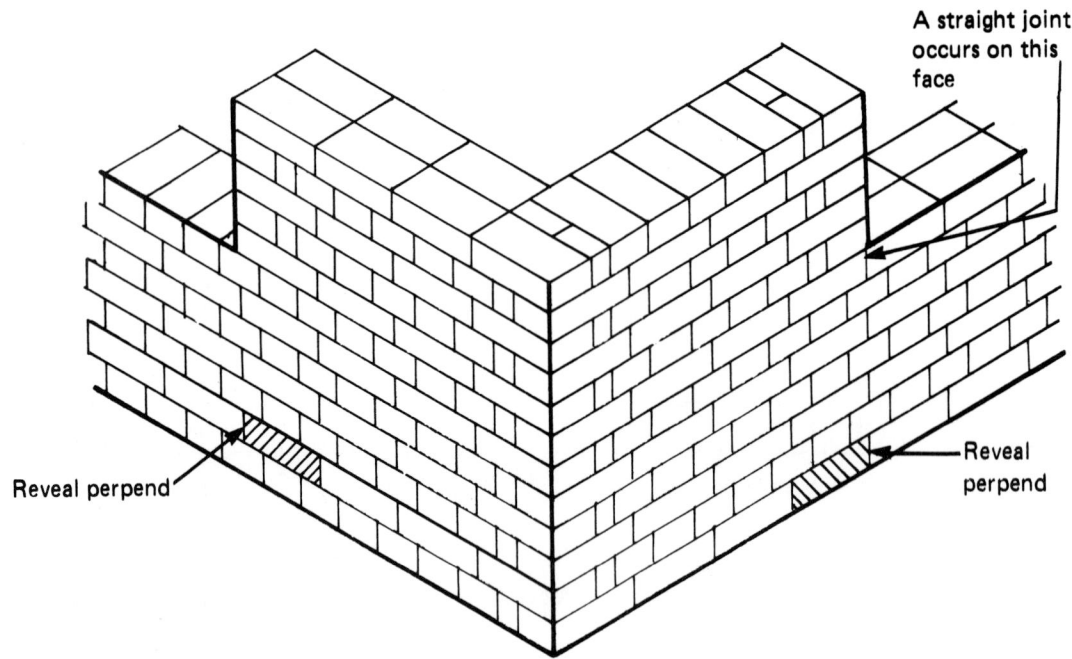

An isometric view of a quoin showing the setting out of face bonds on adjoining faces

Figure 24

Reverse bond in English bond

Reverse bond in Flemish bond

Reverse bond in Dutch bond

Figure 25

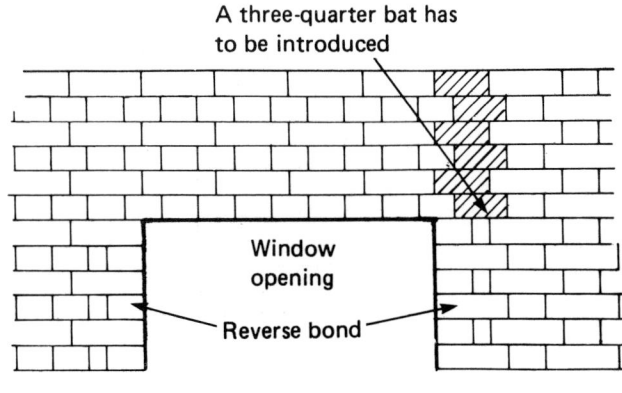

Elevation showing the effect of reverse bond on the bonding at window heads

Figure 26

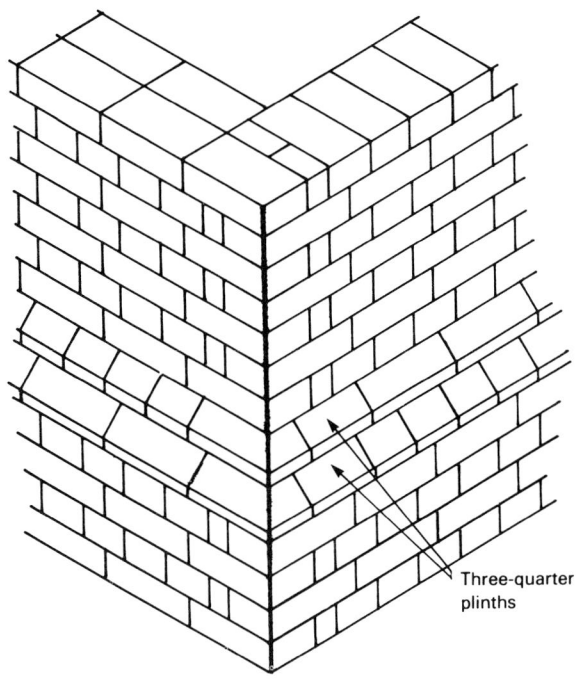

A quoin built in English bond showing the bonding for two plinth courses
Figure 27

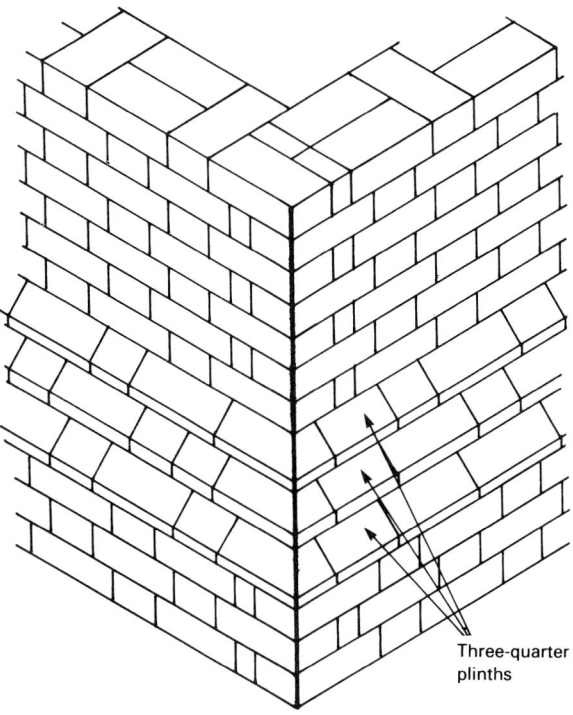

A quoin built in Flemish bond showing the bonding for three plinth courses
Figure 28

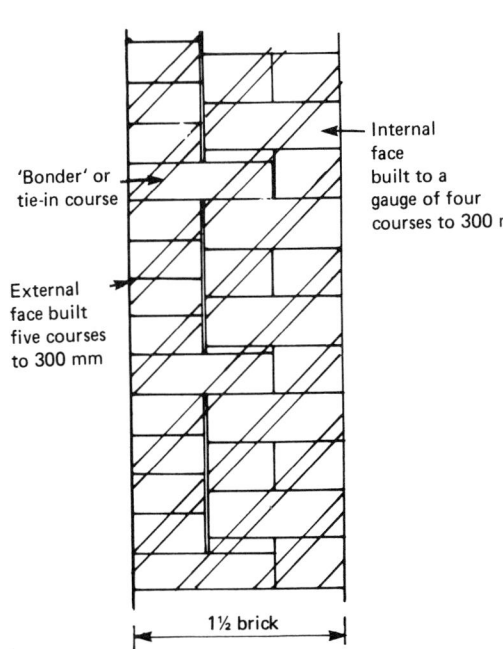
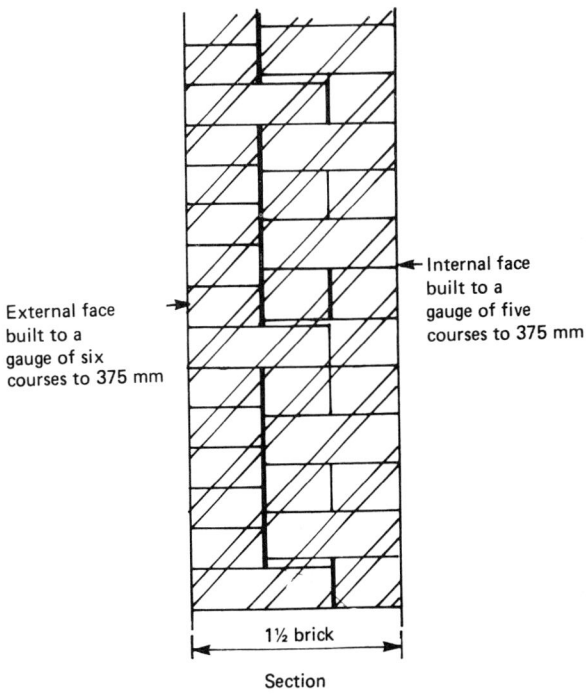

Composite walling using bricks of different thickness

Figure 29

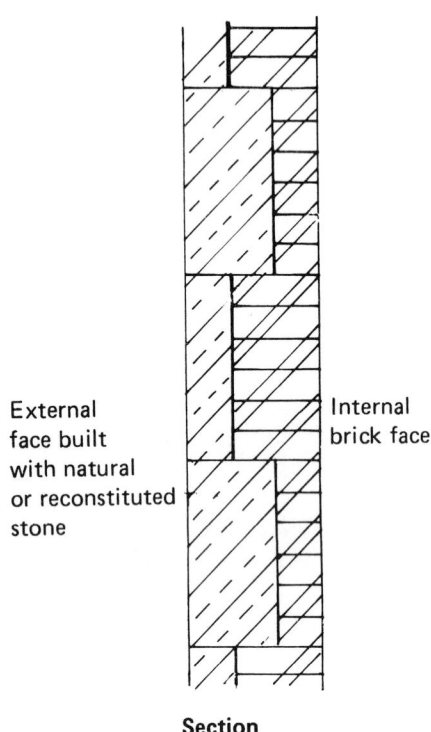

Section

Figure 30

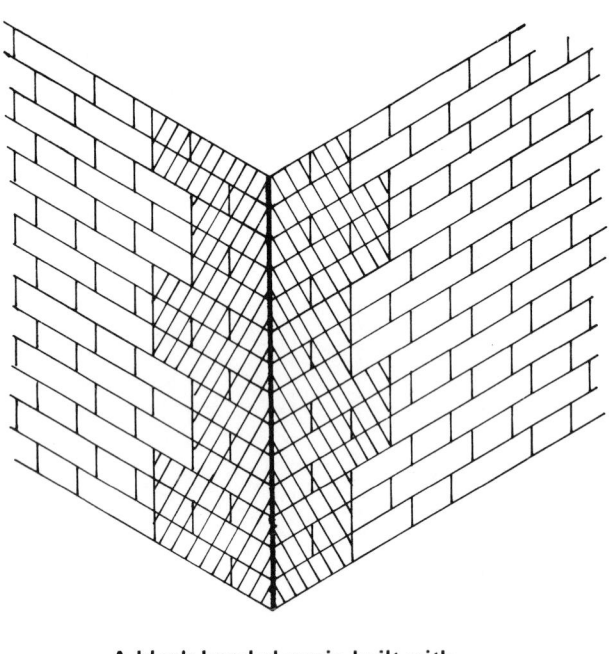

A block bonded quoin built with contrasting coloured bricks

Figure 31

constructed with bricks of different thicknesses. For example the face side of the wall may have bricks of 50 mm in thickness and the inner face of the wall having the traditional bricks of 65 mm. In such cases these walls should be bonded where the courses correspond in height and as frequently as possible without excessive cutting (see Figure 29).

Where stone is used on the face side of the wall with the brickwork backing the work should be bonded together as shown in Figure 30. With this type of composite walling the back of the stone should be painted with a bituminous product to prevent any possible staining being projected through from the brickwork.

Another form of composite walling is encountered when using different bonds on opposite faces. For example one face of the wall may be built with one of the garden wall bonds and the opposite face built with English bond. In these instances there will inevitably be a large number of internal straight joints, so the walling must be bonded across the wall as frequently as possible in order to maintain its strength.

Decorative work

In order to break up the mono-colour of a large mass of brickwork built with only one type of brick, it is common practice to introduce panels of contrasting coloured bricks. An attractive appearance may be produced by using black or dark blue bricks which give a shadow effect and an impression of depth. Such panels may be pointed with a matching coloured mortar which enhances the colour and even greater effect of depth in the elevation of the structure.

Another pleasing use of contrasting colours is to build block bonded corners, as shown in Figure 31.

Interesting patterns in the form of diamonds have been used with great effect in large areas of plain walling in both old and modern buildings. This system has always been popular as a decorative feature as it can be introduced in a wall without involving a great deal of extra expense, labour or wastage. This form of pattern is called diaper bond and an example is shown in Figure 32, but there are many variations using the same theme.

Another interesting way in which a shadow effect may be introduced into a wall is to recess panels of brickwork; these may be built with the same type of brick as is used in the main walling or contrasting colours may be used. If the panels are a half brick in depth then no difficulty is encountered at the bonding of the sills, heads and reveals, except that a weathering should be formed on the sill to shed water. If, however, the recess is to be a quarter brick in depth then the sill and head may be built with plinth bricks and the reveals with cant bricks (Figure 33). This

Diaper bond

Figure 32

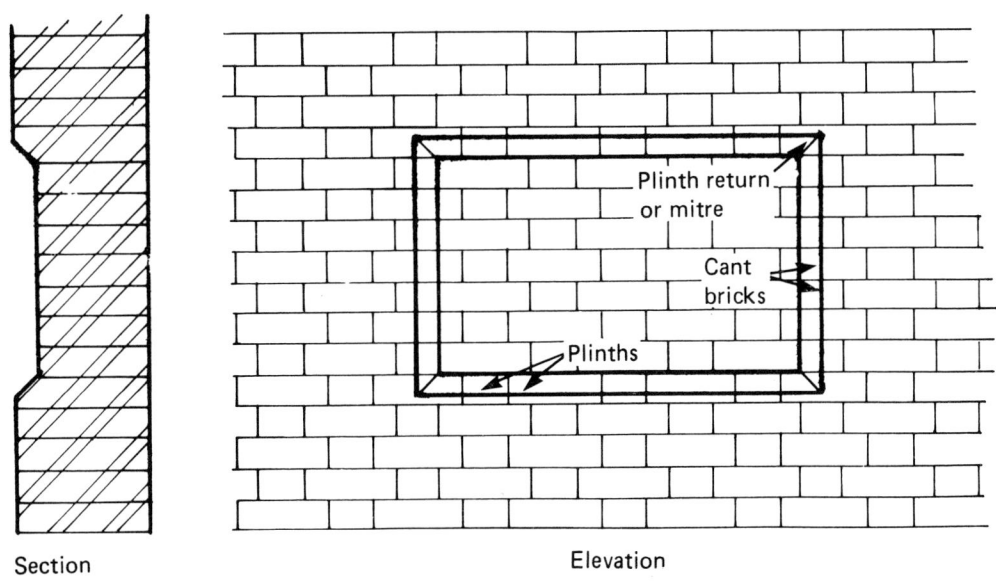

Section   Elevation

The formation of a recess in a wall
using plinths and cant bricks

Figure 33

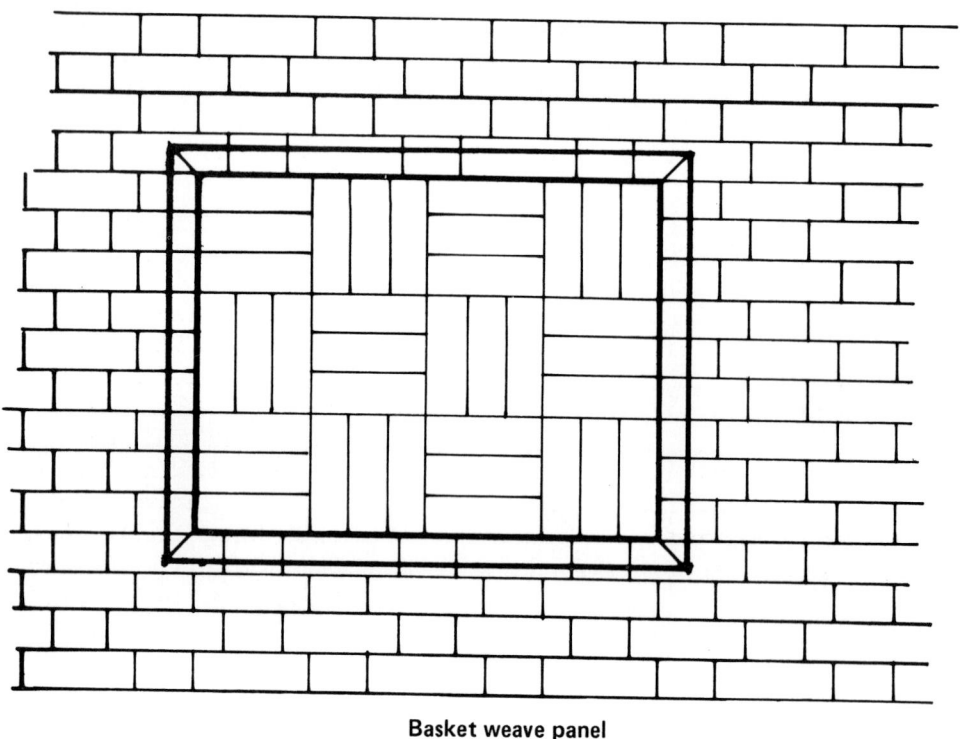

Basket weave panel

Figure 34

A panel built with herring-bone bond

Figure 35

forms an attractive feature and provides a very efficient weathering to the sill.

Sometimes these recessed panels may be infilled with brick patterns such as basket weave or herring-bone bonds as shown in Figures 34 and 35. In the case of the basket weave bond it is essential that the vertical courses are kept truly perpendicular and that each layer of pattern is truly straight and horizontal. With the herringbone bond the panel should be set out on a hard surface and the bricks laid dry within the setting out with the correct allowance between them for the mortar joint, then the outline of the panel marked on the bricks. The bricks forming the edges of the panel can be cut on a mechanical saw.

While any additional labour amounts to extra costs, nevertheless the foregoing illustrates how some decorative features may be introduced into walling without involving exhorbitant additional expenses.

Matching brickwork

The carrying out of repairs and restoration of brickwork or extending an existing structure often necessitates the matching of the new work with the old. A great deal of care is needed when confronted with this situation as the colour of the new work is not constant and will change (generally to a lighter shade) as it dries out. Unfortunately this change of colour is rather indeterminate both in the length of time it takes to change and degree of its intensity of colour, particularly with the mortar. It is therefore recommended that specimen panels be built; the type of brick, and the composition of the mortar used being carefully noted in each panel. These should be left as long as possible so that the changes can be carefully recorded.

When a particular panel has been decided upon the craftsman must ensure that the same type of bricks and sand are used in the work, also the correct gauging of the mortar. To assist in this, gauge boxes may be made and used to measure the ingredients of the mortar accurately. These are usually bottomless boxes as shown in Figure 36. If care is exercised then a reasonably matching mortar may be obtained.

When matching very old walling it may be difficult to find suitable bricks in sufficient quantity to carry out the work. New bricks may look incongruous and one may have to search around for older or second-hand bricks. In such cases it may be possible that another building on the same site or in the immediate locality is being demolished, in which case the possibility of salvaging the bricks should be investigated and, subject to the architect's approval, be used in the repair work. If approval is given then the salvaged bricks must be thoroughly cleaned of all mortar and only the best ones selected for use. This is quite an expensive way of obtaining suitable bricks for special work but it can be very effective and the appearance most rewarding when completed.

Apart from the decorative advantage, if the bricks were hard burnt stocks and have been well weathered, then they may be quite an asset in the rebuilt walling as they will be long lasting and will stand up to the various weather conditions extremely well.

If the bricks are scarce and cannot be spared to build specimen panels then a matching mortar may be found by making up small samples of about 75 mm by 50 mm by 25 mm thick and allowing them to dry. When dried they may be placed in a warm room. Control samples should also be prepared by putting a bed of each mortar between two bricks (damaged specimens of the salvaged bricks will be very suitable for this) and allowed to dry. The colour changes should be carefully noted as mentioned before. This method will give a reasonable guide in a fairly short time to the selection of a mortar mix for the walling, and is far better than trying to use any rule of thumb methods.

When using new bricks for repairs, and after a sample has been approved, it is usual for random samples to be taken from the initial loads and deposited with the architect or his representative on the site to ensure that subsequent deliveries match. The bricks should also be unloaded with care and those which are chipped or damaged rejected.

In dry weather the walling should be dampened to prevent the mortar drying out prematurely. In cold weather the bricks should not be wetted but laid dry. In wet weather, freshly laid brickwork should be protected during interruption through rain, and at the completion of each day's work, by covering it with polythene sheeting weighted down with loose bricks.

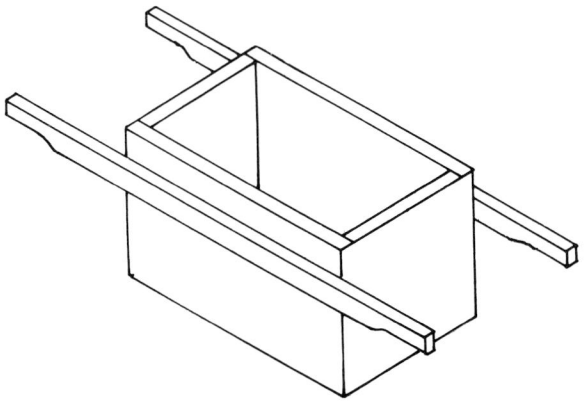

A gauge box

Figure 36

3 Jointing and pointing

In general, brickwork that has been well built requires only the minimum of maintenance work to keep it in a good condition. However, as a building ages the surface of the mortar joints may show signs of deterioration which may be attributed to one or more of the following causes:

1. Water damage from leaky roofs, gutters, rain water pipes and leaking plumbing.
2. Ingress of water due to faulty damp-proof courses at ground floor level, sills, lintels, eaves and parapets.
3. Poor mixing or incorrect proportioning of the original pointing mortar.
4. Frost damage.
5. Sulphate attack due to the presence of soluble salts either in the water entering the walling or in the bricks themselves.
6. Permanent damp situations where the normal evaporation process is restricted due to being sheltered by deep overhangs, or in the close proximity of adjacent buildings.
7. Structural movement.

The causes of jointing decay should be corrected before repointing is undertaken, either by taking precautions against the entry of water or by removing the source of the water.

The amount of repair necessary will depend upon the degree to which the decay or movement has proceeded. In the worst circumstances partial rebuilding may be the only remedy. Careful probing into the bed joints will give a clear indication of the condition of the walling. In cases where the joints have completely deteriorated it is a waste of time and effort just to carry out a repointing repair. For satisfactory results to be obtained when repointing the existing bed joints must be reasonably sound so there will be good adhesion between the bed and the pointing mortar.

Jointing

This is the term given to finishing the joints off as the work proceeds. It is usual to lay a course of bricks then to finish the joints, but this will vary somewhat, according to the weather and the suction rate of the bricks with which the wall is being built. This method of jointing has a distinct advantage in that the mortar in the joints is unified so there is less likelihood of the pointing being forced out by frost. It is also more economical than pointing the work after it is completed.

The disadvantages are that it restricts the use of coloured mortars. Also there is a possibility that the sections of the walling will vary in colour if the mortar has not been carefully gauged. Additionally, if it should rain while the work is in progress then there is every chance

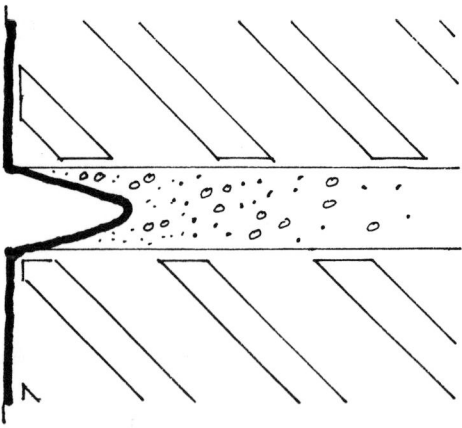

Incorrect method of raking out

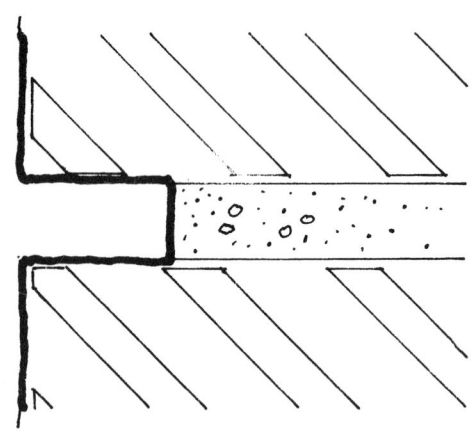

Correct method of raking out

Figure 37

that the joints will be difficult to finish off, similarly, if it happens to rain at the end of the day's work when the operatives have left the site.

Pointing

Pointing is the term given to finishing off the joints after the wall is completed. As each section of the work is finished so the joints are raked to a depth of about 15 to 20 mm. This operation should be done with care and Figure 37 shows the correct and incorrect methods of raking out. It is so often done with a cavity wall tie but in practice this is rather too severe because if the bricks tend to be on the soft side then the ties will take off the arrises, thus causing the joints to be widened. It is, therefore, much better to use a piece of wood which has been whittled down to the thickness of the joints.

After the joints have been raked out they should be thoroughly brushed out to remove any loose particles which may remain in the joints, and also to ensure that the arrises are clean. When the wall has been completed and ready for pointing, it should be thoroughly cleaned down with water and a scrubbing brush used to remove any stains. This brushing should be carried out with care to ensure that no damage is done to the facework. Non-ferrous wire brushes may be used on hard facings to remove obstinate stains but again care must be exercised to avoid damaging the facework.

After the wall has been wetted and cleaned down it should be allowed to dry out so that it is damp enough to receive the pointing mortar. This is to avoid excessive drying shrinkage occurring after the pointing has been completed and breaking down the bond between the pointing mortar and the bed joint, particularly when it is subjected to frost attack. Also it is very difficult to point a wall and keep it clean if it is very wet. Ideally, a wall should be clean, joints well raked out, and in a damp condition to achieve the best results when pointing.

The advantages of pointing after the walling is completed are that the work can be cleaned down thoroughly as the scaffolds are dismantled; better control of the colour of the pointing mortar can be maintained and coloured mortars can be used with remarkably good effects.

Mortars for pointing

These may vary in composition according to requirements, but as a general rule the ultimate strength of the pointing mortar should not exceed the strength of the bricks used in the walling. Hard dense mortars can cause serious deterioration of some types of softer facing bricks. This is because the hard mortar restricts the evaporation of any water through its surface and the bricks can become saturated. If they are subjected to frost attack then this may freeze the water and break down the structure of the bricks, causing spalling on the surface. In severe cases this can result in the bricks being eaten away and the mortar remaining in ridges in front of the brick surfaces.

For hard burnt, dense bricks in exposed situations, such as parapet walls, a fairly hard mortar may be used. For example a cement:sand mix 1:3. But for general facework a cement:lime:sand mix of 1:1:5 or 1:1:6 should be quite adequate, though the eventual colour of the mortar will be a more deciding factor in the selection of the materials and their ratios.

When pigments are introduced to colour pointing mortars several points must be considered. The materials must be carefully and accurately measured, preferably by weight rather than by volume, because of the tendency of the sand to bulk under damp conditions (sand can increase in bulk up to 30 per cent of its dry volume) which may seriously affect the colour and strength of the pointing mortar. Also the pigments must be applied in accordance with the manufacturer's instructions. These pigments are usually very dustlike and when adding it to the mortar mix should be sheltered from strong winds which will blow some of the pigment away and may affect the desired results of the mix. As some sands may vary quite considerably in their colour, grading and composition, then it is essential that the same type of sand should be used throughout the whole of the work.

When selecting a coloured mortar for pointing it is most important that the colour of the mortar is complementary to the colour of the bricks. Some colours can enhance the brickwork while others can 'kill' the wall colour. This is a vital factor and one which so often is not always fully considered too carefully by those responsible for work of this type. A simple experiment may be carried out to demonstrate the effects of colours by building a one-brick wall about 2 metres square and dividing it into four or more sections and pointing each area with a different coloured mortar. The resulting appearance can be most remarkable and cases have been known in which onlookers have been convinced that different bricks have been used to construct the wall.

Very pleasing results can be produced in walling. For example, by pointing all the perpends in a wall with a mortar the same colour as the bricks, thus 'blinding out' all the cross joints and then pointing the bed joints with a contrasting mortar. This gives the effect of long lines of brickwork. Similarly, by using a monk bond to build a wall and 'blinding out' the joints between each pair of stretchers and then using a contrasting mortar for the remainder of the joints, it is possible to give the effect of elongated stretchers. This is called Flying Flemish Bond.

Procedure for pointing

In order that the bed joints maintain a continuous line across the wall it is usual practice to point the perpends first followed by the bed joints. As each section of work is finished it should be brushed down lightly with a soft brush.

Types of pointing

Flush pointing

This is a common finish that is used on all fair faced work,

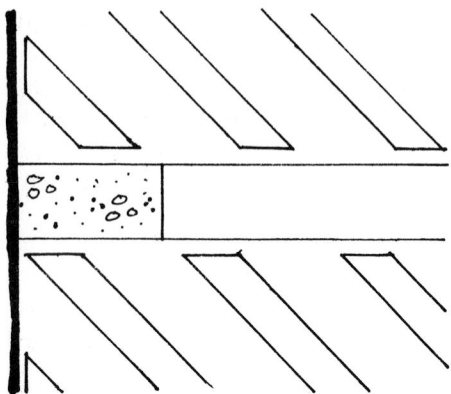

Flush joint

Figure 38

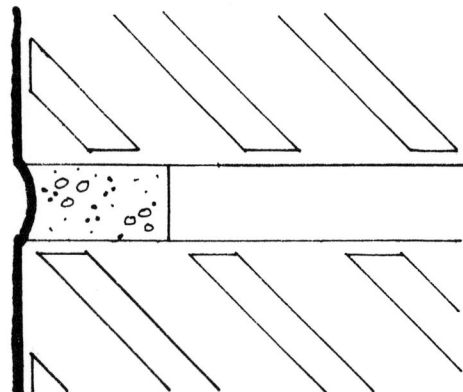

Round or tooled joint

Figure 40

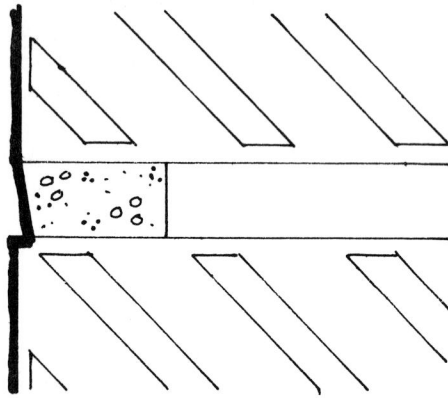

Struck joint

Figure 39

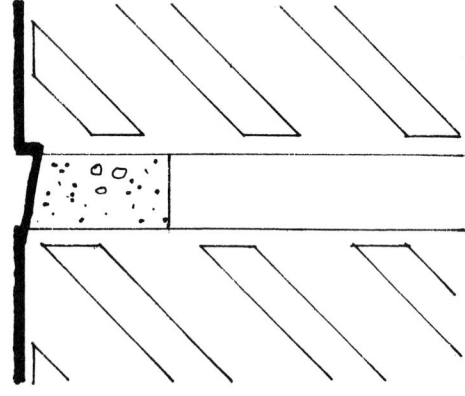

Weather struck and cut joint

Figure 41

internal and external. The joints are filled flush with the brick surfaces thus forming no lodgements for the collection of dust (Figure 38).

It is a very effective method of pointing particularly in conjunction with hand-made sand faced bricks, but it is one which is not too easy to apply as it requires much care in keeping the joints really full and free of 'pinholes', and also the faces of the bricks clean. The joints are filled then usually rubbed over with a piece of hessian or sacking, which must be changed frequently in order to avoid mortar being smeared over the brick faces.

Struck joint

This is used mostly on internal fair faced brickwork and is applied by running a trowel along the bed joints while pressing it in on the lower edge of the joint (see Figure 39). The perpends are also pressed in on one side of each joint. This type of pointing is not recommended for external

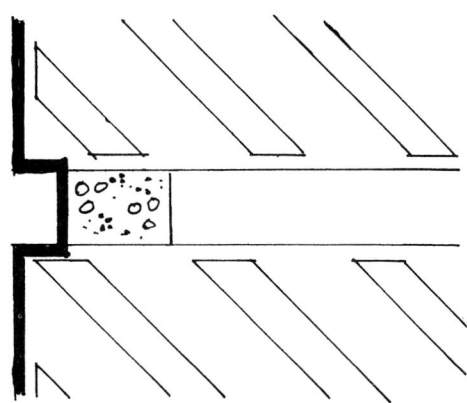

Recessed joint

Figure 42

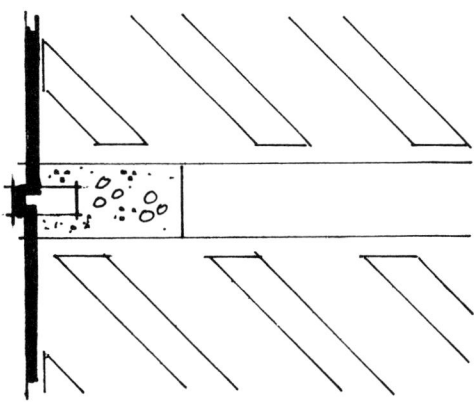

Tuck joint

Figure 43

work as there is a possibility of water collecting on the upper arrises of the bricks which, if subjected to frost attack, could damage them.

Rounded or tooled pointing

Sometimes this is referred to as 'bucket handle pointing'. It is a simple but very effective method of finishing off brickwork joints which are filled with the pointing mortar then rubbed in with a jointing iron, as shown in Figure 40. When finished, the work should be lightly brushed over with a soft brush. This pointing finish has the advantage that the mortar is well pressed into the joints and it looks extremely good when used in conjunction with bricks which are regular in size and shape and the bed joints even in thickness throughout the walling.

Weather struck pointing

This is recommended for use in external face work as it sheds the rain water from the joints thus giving protection to the bricks. Also, if there should be any irregularities in the bricks or joints, the pointing can be cut on the lower edges so that all the joints appear to be the same thickness. With this method the pointing mortar is applied with a trowel and pressed in on the upper edge of each joint (Figure 41) and the lower edge cut off with a trowel, or frenchman, and pointing rule. The perpends are filled before the bed joints and cut off with a trowel. Another advantage with this method is that it forms shadow lines on each of the bed joints and gives 'depth' to the wall face, giving it an interesting appearance.

Recessed pointing

This is not commonly used for external work unless the bricks are hard and dense, but it is an excellent method of finishing off internal walls built with face bricks, brick fireplaces and so on. The recessed joint is formed by rubbing in the joints with a jointing tool or a short length of wood which has the same thickness as the brick joints (Figure 42), then brushing lightly with a soft brush when completed.

Tuck pointing

This is an expensive method of pointing and is only likely to be used on very old buildings. It is particularly useful when applied to arches and similar features in order to give them the appearance of having been built with gauged work.

There are two methods of tuck pointing. One is called Bastard tuck pointing in which the joints are raked out, brushed clean and filled with the pointing mortar which is left protruding slightly from the face of the bricks. The mortar is then cut off on both sides of the perpends and top and bottom of the bed joints with the aid of a frenchman, or trowel, and pointing rule (Figure 43). By using this method the craftsman is able to hide any irregularities in the joints and chipped arrises which may have been caused when raking out the old joints.

The other method of tuck pointing is carried out by first filling all of the raked out joints so that they are flush with the bricks, and then colouring the whole of the walling with a compound of copperas. This is mixed with a colouring agent and rubbed well into the walling. White lime putty is applied to the joints with the aid of a tuck jointer and pointing rule. The edges of the joints are trimmed off with the aid of a frenchman and pointing rule. A variation of this method of pointing is to 'blind out' all the joints with a mortar matching the colour of the bricks and then apply the lime putty as before. Tuck pointing is very effective but requires much care and skill in its application.

When using lime as a pointing material it should be mixed thoroughly with water in a large container until it is like a thick cream, and left for as long as possible (allow a minimum of 24 hours) before using. This will allow the lime to slake thoroughly and avoid any possible chance of quicklime being used in the pointing mortar.

Additives

When carrying out restoration work the use of additives should generally be avoided. When working in cold weather there may be a tendency to use a calcium chloride or an admixture containing a percentage of calcium chloride as an anti-freeze agent. This is a deliquescent salt which will attract moisture from the atmosphere, and is liable to cause serious deterioration in the joints.

While some additives, such as plasticisers, may be harmless, nevertheless many additives are comprised of soluble salts which can cause efflorescence on the face of the brickwork, and, in sufficient quantity, can also have harmful effects on the bricks. Approval should, therefore, first be obtained from the architect before any are introduced on the site. In fact it is usual for specifications to state quite clearly that additives must not be used on restoration work.

4 Cleaning brickwork

When a structure is in need of restoration it is most important that the walling is closely inspected, with full details of the extent and nature of the repairs to be carried out carefully noted and recorded. If the walling is covered in plant organisms or stains then it is essential that the work is thoroughly cleaned down so that a complete appraisal of the work to be done may be made.

Generally it is sufficient to clean down brickwork by wetting and scrubbing with bristle brushes or, for more obstinate stains, with the aid of non-ferrous wire brushes. However, extreme care must be taken to ensure that the surfaces of the bricks are not scoured. Ferrous wire brushes may be used with care on very hard bricks, or in extreme cases of severe staining.

The cleaning down process should be commenced at the top of the walling working across in horizontal stages and working down the face of the wall. When the walling is cleaned down a full assessment of the extent of the necessary repairs may be carried out. It also gives an opportunity for the wall to dry out somewhat before such repairs are commenced.

Removing plant growths

Algae, lichens and mosses are frequently found on external walls of many structures particularly those which are very old. In the UK moisture conditions are generally more favourable on northern aspects than on south facing walls, though lichens will often flourish in southern aspects. In some cases the growth may be quite luxuriant especially where the owners have desired the pleasing effect and have even encouraged it by applying washes of cowdung and water. When such growths need to be destroyed they should be removed as far as possible with a knife blade, spatula, stiff bristle brush or non-ferrous wire brush, ensuring that the surface is not scoured. The walling may then be sprayed with a dilute fungicide. A pneumatic type garden spray is very suitable for this work as it will give sufficient pressure at the wand nozzle without causing 'bounce back' and spray drift.

A flood coat should be applied commencing at the top of the vertical surface to be treated and moving across horizontally and slowly to allow about 100 mm run down. The next horizontal pass to be made across the previous run down. The treated area should then be left for several days and the dead growth brushed off with bristle brushes. The last operation is usually more effective when the dead organism is dry.

A further toxic wash such as household bleach, which is very effective against many organisms and has no active residue, or one of the marketed specialised solutions, such as Thaltox Q or equivalent, may be applied in the same manner as the fungicide. It is better to use separate sprays for the two operations. If the same sprays are used then they should be thoroughly cleaned out before changing from one solution to the other. The surface should be allowed to absorb the solution and then a second application made if required. If mould growths are found on internal walls this is a sign that there is rain penetration or defective damp-proof courses, then toxic washes may still be applied but extreme care must be taken when using them in enclosed spaces.

All of these toxic solutions must be handled with great care as many of them are caustic in concentrated solutions. Even solutions which are diluted ready for use should, if splashed on the skin, be washed off immediately with soap and water. Particular precaution should be taken to ensure that they are not splashed into the eyes. Protective clothing and goggles must be worn by the operatives.

If an area of skin larger than about 30 cm^2 should be contaminated by any toxic solution, the person affected should be taken to the nearest casualty ward without delay and the hospital staff informed of its nature. All suppliers issue detailed instructions concerning the handling of these materials and these should be followed meticulously. A formalin based toxic solution should not be used in an enclosed space as this can be a health hazard unless good ventilation is arranged.

Some types of toxic washes, such as the copper, carbonate, ammonia solutions, are very efficient in use but may stain some materials. Therefore, it is wise to test on a small area before using on the main walling. There may also be a tendency to use dilute solutions of hydochloric acid or muratic acid to remove obstinate stains, but indiscriminate use of such materials is ill advised and it is far better to rely on the methods which have been previously described.

Rendered walling

In some very old buildings it is quite common to discover face brickwork which has been subsequently rendered. If this rendering is to be removed and the face brickwork restored, then the rendering coat should be gently hacked off. Extreme care must be taken to ensure that the face work underneath the rendering is not damaged.

If the rendering is of a comparatively soft nature then the hacking off may be reasonably easy, but the cleaning down of the walling will require considerable care. If the staining from the rendering is dry then it should be possible to remove much of the staining with a dry bristle brush. After this the walling should be cleaned down with water and brushing as previously described.

Assessing the work to be done

Once the surface of the walling has been exposed and

cleaned thoroughly the nature of the defects can be assessed. Such defects may include the following:

1. Serious deterioration of the bed joints and possibly the bricks due to:
   (a) continuous wetting;
   (b) sulphate attack.
2. Spalling of the brick faces by:
   (a) frost action;
   (b) too dense a mortar being used for bedding the bricks or pointing them.
3. Serious deterioration of chimneys due to sulphate attack.
4. Cracking in the walling due to structural movement caused by:
   (a) uneven loading in the structure;
   (b) uneven load-bearing capacity of the soil which may be caused by changes in ground water levels or site drainage;
   (c) settlement of the foundations caused by trees growing too near the structure, particularly when they have been planted too near the building and the foundations of the structure are not very deep;
   (d) subsidence due to excavations or mining works adjacent to the structure;
   (e) frost heave in chalk sub foundations;
   (f) vibration caused by heavy traffic;
   (g) thermal movement;
   (h) drying shrinkage.
5. Corrosion of embedded iron or steel.
6. Unstable cavity walls due to badly corroded wall ties.
7. Damage to corbels, string courses, sills, arches and similar feastures by settlement or erosion.
8. Deterioration in parapet walls, coping stones and badly corroded metal cramps.
9. Dampness penetrating to the interior of the building due to:
   (a) defective damp-proof courses;
   (b) blocked cavities;
   (c) defective cavity wall insulation.
10. Bulging and unstable walling.
11. Flood damage.
12. Sagging of ground floors and partition walls due to inadequate consolidation of the infilling under the solid flooring.
13. Effects of fire.

Recognising the defects

Sulphate attack
For this to occur in brickwork three elements must be present; water, soluble salts and alumina, the latter being found in Portland cement and in many hydraulic limes. If only two of these items are present then no trouble is likely to occur. Thus if a wall containing soluble salts is built with Portland cement and is kept dry then it will remain in perfectly good condition. If, however, the wall becomes wet and remains so for long periods, then it is very liable to sulphate attack.

When this happens the affected mortar is generally found to be cracked along the length of the joints and the surface of the joints fallen off. The joints will expand and in severe cases may be reduced to a soft mush. Mortar affected by sulphates has a whitish appearance. Where it is in contact with the bricks it is usually whiter than in the centre of the joint. The height of the walling can be increased quite considerably due to the expansion of the joints, and, in the case of affected chimneys, the weathered side may be more affected than the sheltered side, possibly causing severe tilting. The bond between the bricks and the mortar will also be broken down, so such affected chimney stacks are potentially dangerous. Also, in cavity walls any expansion upwards of the outer leaf can cause the wall ties to tilt so that there is a possibility of rain penetration to the inner leaf.

Where the external walls are rendered with a Portland cement mortar, sulphate attack will cause a considerable amount of cracking, usually horizontal, and mainly along the line of the mortar joints. Sometimes the rendering may become detached from the brickwork depending on the extent to which the under surface has been attacked.

Permeable renderings are less liable to cracking as they permit evaporation from the brickwork than more dense mixes, and the less water trapped in the walling means that the less is the likelihood of sulphate attack.

Frost attack
This has a similar appearance to sulphate attack, but careful inspection will usually enable the person carrying out the inspection to distinguish one from the other, as frost attack usually has less cracking along the joints and there is generally more spalling of the brick surfaces.

Damage due to the use of dense mortars
This is readily distinguished from frost attack because in very severe cases the bricks may be eroded to such an extent that the mortar remains hard and may actually protrude from the surface of the bricks.

Cracking in walling
This is easily identified, but the cause is not so easily diagnosed and generally requires a more extensive survey in order to determine the factor or factors bringing about

the defects. Such surveys should include investigating:

1. If excessive loading has been superimposed on the walls of the structure.
2. The condition of the soil under the foundations and the level of the groundwater. Also note if any excavations have been carried out adjacent to the structure thus causing the area underneath and surrounding it to be drained, which could affect the load-bearing capacity quite seriously.
3. The depth of the foundations. If they are shallow and have been built upon chalk then any movement upwards could be caused by frost heave. The deeper the foundation level then the less is the likelihood of this occurring.
4. Drying shrinkage which is not all that common in brickwork but is likely to be encountered in blockwork.
5. Thermal movement. This may happen where a wall is restrained at both ends without provision for movement. Fluctuation in temperature may set up stresses sufficient to cause fracture or distortion of the brickwork.
6. Intensity of traffic. The structure may be subjected to excessive vibration due to traffic volume or the actual size of vehicles, which may have increased over the years.

Very often the pattern of the cracking can give clues to the causes. For example, if the crack travels more or less vertically in the walling, this generally indicates stresses created by thermal movement or excessive loading. If the cracking travels in an inclined direction it would suggest a lifting or sinking action and the foundations should be examined for defects.

Corrosion of iron or steel
This is readily visible as it usually creates bad staining and spalling of the material immediately surrounding the corroded metal.

Corroded wall ties
In cavity walls where the wall ties have become badly corroded the outer leaf may become unstable and slight movement easily detected. In severe cases there may be patches of staining on the face of the brickwork due to the corrosion of the ties, but this is not so common.

Damaged corbels etc.
Defects in any projecting work such as sills, string courses etc., may be easily seen, but the main points to note when carrying out an inspection is the type of bricks which were used in the construction, and the amount of exposure to which the damaged items are subjected.

Exposed walling
Deterioration in parapet walls, copings, retaining walls and free-standing external walls is likely to be accentuated because of their severe exposure to the elements. Water can reach both sides of the walls and often from the top. Moreover the walling receives no warmth from the interior of the building, therefore its drying is controlled by the weather. Such exposure tends, therefore, to increase any risk of sulphate attack. This also applies to the brickwork in manholes and other underground work.

Dampness in buildings
This may cause severe staining on wall surfaces, may also encourage the growth of mosses, and cause deterioration in the rendering or plaster on the surface of the walls. Such dampness may also be a prime cause of the development of dry or wet rot.

The penetration of water may be directly due to a defective damp-proof course but it can also be caused by mortar falling inside a cavity and not being cleaned out when the work was completed. If such debris collects above the horizontal damp-proof course then this will allow the passage of water from the ground to the internal walling above the damp-proof course. The warmth of the interior will accelerate the movement of this moisture as it will always travel to where the evaporation rate is the greatest. Thus, ironically, when damp conditions prevail, the warmer the interior of the building, the greater will be the tendency for the wall to remain damp.

Dirty wall ties are also a cause of water penetration and these may easily be identified by the isolated patches that will be seen on the interior wall surfaces.

Cavity insulation which has been badly installed may also be a serious cause of water penetration. This usually occurs in isolated patches.

Bulging walls
Such defects may be the result of excessive eccentric loading being placed upon the walling or undue settlement in the foundations. Once again careful investigation must be made before deciding on the extent of the repairs to be carried out to remedy the faults.

Flood damage
Obviously the walls will be saturated but not so obvious is the amount of mud that may be deposited in the cavities, air ventilators to underfloor space and boilers, service access pits, underfloor spaces and spaces behind wall linings.

Damage to foundations may occur during the drying

out of the subsoil especially if they are situated in shrinkable clays. These swell when wetted and the subsequent shrinkage upon drying may cause cracking in the walling. Drains may also become blocked due to the build-up of mud and silt.

In the case of flooding by sea water there may be additional problems created by the walls being contaminated by deliquescent salts which remain permanently damp in very severe cases. Apart from the difficulty in overcoming this problem there is also the possibility of corrosion in fastenings, electrical installations, etc.

Subsidence
The interiors of many houses have suffered serious defects due to the subsidence of the hardcore filling below the solid concrete ground floors; especially where such infill has been excessive and has been inadequately consolidated. In such a case the floor will subside and cracks will appear in the floor surface. Partition walls will also sink and cracks are likely to appear usually in a diagonal direction and often over the door openings. Door frames may become distorted and the doors difficult to open and close.

Effects of fire
These will vary according to the thickness of the walls which have been subjected to the fire, the type of bricks used in the walls and the intensity of the fire. In severe cases cracking and bulging of the brickwork is to be expected. There may also be excessive spalling of the bricks due to the sudden quenching of the walling with water in the course of firefighting. Clay bricks are rather less likely to suffer damage than concrete or sand lime bricks.

Where structural steel members are built into the walling there is likely to be severe distortion due to the amount of expansion that will occur.

The high temperatures reached in a fire will cause gross expansion of beams, roof trusses or lintels built in or resting upon the brickwork. This is likely to displace the brickwork and cause severe cracking.

Recording defects for repair and restoration
Having cleaned down the walling and identified the defects the next step is to record them and the work necessary for their repair or restoration. This must be done in such a manner that all those concerned with the repairs or renovations are conversant with the work to be done.

Such recording, therefore, must be carried out systematically so that every item is included in the details. On small works it is a comparatively easy matter to mark on a drawing the work to be done, to write a simple schedule of operations and to estimate the quantities of materials and labour that will be needed to perform the work.

When restoration work is being carried out on large complex structures, however, it is most important to identify each item of work to be done in close detail and is also essential when tenders are to be invited from contractors to carry out the restoration work.

Plans of each floor are drawn showing the thicknesses of walls, positions of cross walls and partitions, positions of openings and all other relevant details. Also elevations of each side of the structure are produced. These may be obtained by 'taking off' or measuring the building and drawing the elevations from the measurements. Alternatively, a more suitable method is to photograph each face of the building and to produce photogrammetric elevations. This has been one of the most important and progressive advancements in techniques for surveying building facades in recent years. When scaled drawings of elevations of large buildings have to be made showing extensive and complicated details of repair work to be carried out *photogrammetric surveys* may be well recommended because of their great accuracy and speed of preparation compared with hand drawn measured surveys. Most of the research and development work on this system of preparing corrected photographs of building elevations so accurately that they can be used for making scale drawings has been done by the York Institute of Advanced Architectural Studies in conjunction with the Royal Commission on Historical Monuments and by the York Photogrammetric Unit.

Ordinary photographs are clear, attractive and instantly recognisable pictures of buildings, but scale drawings cannot be made accurately from them as they are in perspective and are, therefore distorted, as no lines are parallel. To illustrate this more graphically, if one stands between two railway lines and photographs the track the two rails appear to meet in the far distance. To overcome these difficulties specially designed sterescopic cameras have to be used with auxiliary equipment and special films.

These surveys may now be fed into a hand held site computer which is linked directly into the cameras. The Moss Site Measurement Module is a suite of programs based on the Husky Hunter field computer for survey data collection. Observations may be recorded directly from most types of electronic surveying instruments or entered manually where appropriate.

The operation of the photogrammetric system requires much study and skill and those who may be interested in this subject may obtain excellent descriptive papers from the York Institute of advanced Architectural Studies. Also further information regarding the computer system may be obtained from Moss Systems Ltd., Barclays House, 51 Bishopric Horsham, West Sussex, RH12 1QJ.

Another excellent time-saving procedure is to make use of large photographs (known as *rectified photographs*) prepared to a predetermined scale of the whole of the defective areas. Every detail is then clearly shown with its defects apparent. This enables a close and detailed study to be made and each defect can then be marked off on the photographs for attention. The required repair work can also be clearly indicated and may be readily understood by all concerned.

A measuring rod should always be included in the photographs and if actual site measurements of individual units such as openings, arches, niches and so on can be indicated then this will enable reasonably accurate scale drawings to be made of the elevations.

When carrying out work of this nature it is important to avoid unusual angles of view or tilting of the camera which can cause variations in perspective and consequent inaccuracies of scale. Also if too large an area is attempted in each picture there is likely to be distortion by the camera lens. This work may be carried out in either colour or black and white photography. The use of either Photogrammetry or Rectified Photographic Surveys ensures that everything is included on each drawing and no details are omitted.

From each detailed elevation a skeleton elevation, or key drawing, is made and on these are shown the positions of doors, windows, niches etc. numbered logically and systematically. It is of little consequence whether the numbering starts at the top or bottom of the drawing so long as it is clearly understandable and free from ambiguities.

A common method is to start at the front of the building and number the windows from left to right in all elevations commencing at the basement then continuing with all the windows on the ground floor followed by the first floor, second floor, etc. The window openings would be numbered W1, W2, W3, W4, etc. until every opening in the building is identifiable.

The same procedure would be used for the door openings. These could be marked as D1, D2, D3, D4, etc.

Niches may be marked as N1, N2, N3, etc.
Panels may be marked as P1, P2, P3, etc.
Chimneys may be marked as C1, C2, C3 etc.

Thus any work which is required to be done at any of these features may be readily located and easily identified.

Sections may also be drawn on the key elevation to show any special feature such as recessed reveals, splayed reveals, moulded bricks and so on, but, in addition to these, large scaled drawings would be made of all the mouldings, reveals, cornices, string courses etc. stating quite clearly to which the details refer.

Hatching can also be used with advantage to indicate items of work such as 'work to be demolished' and 'work to be reconstructed' etc., and colour coding may be used on the key drawing to items of work which are similar in operation but at different points in the elevation.

Each item of repair work or restoration is indicated clearly on each key elevation and every one is numbered. If the elevations are easily identified then the work items may be numbered 1, 2, 3, 4, 5, etc. on each elevation. Each item would be indicated by a square or rectangle showing its locality on the drawing. All of these items would be then scheduled and, for easy reference, a schedule may be listed on the key drawing. This schedule would, of course, embrace all trades – bricklayer, joiner, plumber, stone mason etc. – and refer to any particular opening or panel, if such is being worked on. It would also refer to large scaled detailed drawings.

Figure 44 shows an elevation of a building which is to be restored. The numbering of windows etc. have commenced at the basement and carried right round the building then continued on the ground floor, then the first floor etc. So the numbering is hypothetical. A number of features have been introduced to illustrate typical clauses which may be found in the schedule which refers to this elevation.

Note: The following would normally be written on the key drawing.

West elevation

Schedule of works, key and diagram

1 Remove chimneys C1 and C2 down to the horizontal damp-proof course and inspect the brickwork below for soundness: the bricks to be salvaged for re-use. Replace the damp-proof course, soakers, apron, cover flashings, back gutter and re-build the chimneys as per the original, using, as far as possible, the bricks salvaged from the demolition. Where new bricks are required these should be supplied by .... Co. Ltd. The brickwork to be built with a sulphate resisting cement, lime and sand mortar in the proportion of 1:1:6 then raked out and pointed to match the main walling.

2 Remove the coping stones and the brickwork to the parapet wall which has been affected by sulphate

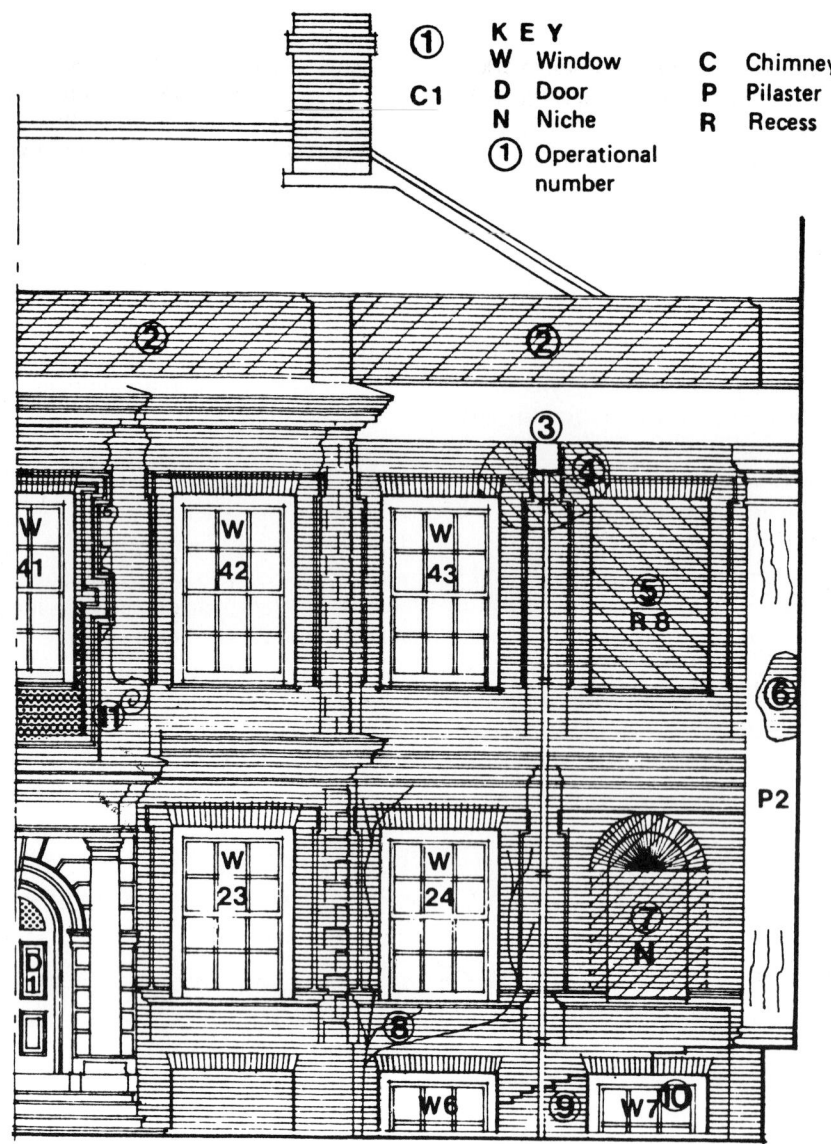

West elevation · key diagram (part)

attack. Salvage as many bricks as possible for reuse.

The box gutter behind the wall to be replaced and any defective timbers underneath and at the outlet to be cut out and replaced, the gutter to be made watertight. Also a 225 × 225mm outlet to be provided throughout the parapet wall at the rainwater head and to be covered over with a Welsh arch.

The parapet wall to be rebuilt, using the salvaged bricks on the face side of the wall and new bricks a.b.d. for the chimneys to be used on the back face. The wall is to be built with a sulphate resisting cement:lime:sand mortar in the proportion 1:1:6, then raked out and pointed with a mortar to match the existing walling.

The feather edged coping to be inspected for soundness and replaced, by bedding soundly in cement:lime:sand mortar a.b.d. for building the parapet wall and jointing with non-ferrous cramps. Any defective stones to be replaced with new stones cut to match the existing.

3 Remove the defective rainwater hopper and replace with a new lead hopper, dress the new box gutter lead around and ensure watertightness. Inspect the lead rainwater pipe and replace where necessary.

4 Rake out all joints to the brickwork surrounding window No. W97 until sound mortar is reached and brush out thoroughly. Carefully remove the gauged camber arch and at least one course above. Install an arch centre over the opening, build up the skewbacks and reset the arch using a white lead and shellac as the matrix. Replace the bricks above the arch ensuring that the existing work is well underpinned, rake out and repoint all the brickwork surrounding

the opening to match the surrounding brickwork.

5 Carefully hack off all the mortar rendering in the recesses R9, R10, R11, R12, R13, R14 and R15, and clean down the brickwork and repoint with a mortar to match the surrounding brickwork. Any damaged bricks in the recesses to be carefully cut out and replaced with brick slips not less than 50mm thick.

6 Hack off rendering to pillaster No. P2 and re-render with a mortar consisting of sulphate resisting cement:lime:sand to match the pillaster P1. The mouldings to be in accordance with the full size details shown in drawing No

7 Fill in all the joints in niches N1, N2, N7 and N8 with a lime putty or white lead and shellac. Carefully replace all damaged bricks with suitable red rubber bricks cut to shape and clean down the whole with a very fine carborundum stone or abrasive cloth or paper. Avoid the use of any harsh abrasives to prevent any damage being caused to the surface of the bricks. The hard skin on the face of the bricks is to be maintained as far as possible.

8 Carefully remove the ivy from the area surrounding windows W1, W2, W32 and niche N1. Also remove algae and licens from the surface of the bricks with the aid of a spatula, knife blade or bristle brush ensuring that the surface of the bricks is not scoured. Spray the affected area with a dilute fungicide by using a garden spray. Apply a flood coat at the top of the surface to be treated and move across horizontally allowing about a 100mm run down at each pass and continue until the whole of the wall area has been treated. Allow to dry then brush off the dead organism with a bristle brush. A further solution of household bleach should then be applied in the same manner as for the fungicide. When dry the walling should be repointed, where necessary, to match the surrounding walling.

9 Remove the brickwork surrounding the window opening W6 down to the damp-proof course. Remove the damp-proof course and replace with a lead lined mineral felt ensuring that there is ample overlapping of the existing damp-proof course to avoid any penetration of damp. Re-build the brickwork, carefully pinning up the existing brickwork, raking out and repointing to match the surrounding walling a.b.d.

10 Remove the damaged gauged camber arch and set out and cut a new gauged arch to match the existing arches, using red rubber bricks and set in white lead and shellac. The surrounding brickwork to be rebuilt and pointed to match existing.

11 The brickwork in the portico to be carefully raked out and repointed with a tuck pointing by filling in the existing joints with a mortar to match the colour of the bricks and then applying a white cement mortar with a tuck jointer, the joints to cut off on both sides to a width of 6mm.

General

Any open joints in the face brickwork are to be filled in with a matching mortar and the walling to be cleaned down with the aid of a bristle brush and clean water as the scaffolding is dismantled.

5 Repairing brick facework

Remedial work

Brickwork which has been well constructed with good quality materials, has ample protection against the weather and is not affected by other influences will give excellent service for many years. Buildings which have been constructed in brick can, however, suffer damage through accident, long neglect, also as a result of deficiences in design, materials or workmanship. In such cases it is essential that before attempting to carry out any repairs, remedial work must be done and, in some cases, special precautions must be taken during repair work to prevent recurrence of the trouble. For example, if a wall has been continually saturated through a leaking gutter or down pipe, then repairs to the plumbing work must be completed first before starting work on the brickwork.

The following recommended procedures have been listed in the same order as the 'recognising of defects' in Chapter 4.

Bed joints damaged through continual wetting

Continual wetting of a wall due to defective gutters etc. may only result in softened joints which may also be badly eroded because of the softening process. In such cases, after the remedial work has been completed, the joints should be raked out to a depth of at least 15-20 mm, or until a firm bed is found, and then the facework repointed to match the existing general walling, taking every care in the matching etc. as described in Chapter 3. It is also good practice to extend the repointing to an area outside of the affected patch in order to merge the repair into the general walling and disguise the repointing as much as possible.

In instances where the bricks themselves have become damaged through continued wetting, these should be carefully removed, preferably by cutting out the joints with a plugging chisel or using a masonry drill, or both, and trying to remove the bricks whole. This is recommended because it may be difficult to obtain good matching bricks and, although the brick may have spalled on one face, it may well be possible to re-use the brick by reversing it and building the spalled surface into the interior of the wall. Where replacement bricks are readily available, then the defective bricks may be cut out with a hammer and chisel. It is important to keep the chisels sharp in order to avoid damaging the bricks adjacent to the defective bricks being cut out, and to prevent the wall from being subjected to undue vibration by unnecessary hammering.

When the defective bricks have been removed the voids should be brushed out to remove any loose dust and the replacement bricks carefully bedded and rebuilt into the wall ensuring that the joints are kept full, especially the last joint which may be underpinning the existing work. This joint should be compacted by using a piece of wood cut down to the same thickness as the joint and pressing the mortar home. Care should also be taken to ensure that the replacement bricks are bedded in line with the courses in the existing work so that the repair work is as inconspicuous as possible and eventually merges into the existing walling.

Sulphate attack

In mild attacks these may be treated in a similar way as that described for damage due to continual wetting, but in severe cases, however, there may be deformation of the brickwork due to the expansion of joints. The amount of repair work necessary will depend upon how far the action has gone, but in any case the action will be liable to continue unless the brickwork can be dried out and kept reasonably dry, too.

As a general guide, where rebuilding is necessary in badly damaged walls, the replacement bricks should have a low sulphate content (if possible) and the mortar be comprised of a mix of sulphate resisting cement, hydrated lime and sand, in the proportion of 1:1:6, and pointed with a mortar to match the existing walling. If cement is to be a component of the pointing mortar then sulphate resisting cement should be used instead of Portland cement.

If a wall has been previously rendered and this has been subjected to cracking due to drying shrinkage, then this may have encouraged sulphate attack in the rendering and the joints due to the water being trapped in the interior of the wall. In these cases the rendering should not be patched but stripped off, and the wall allowed to dry. It is then preferable if the wall is then cleaned down, as described in Chapter 4, and re-pointed. If, however, the wall is to be re-rendered a mix of sulphate resisting cement, hydrated lime and sand (1:1:8 or 1:1:6, preferably the former) should be used.

Damage caused by frost action or using too dense a mortar

Where the joints have been damaged without disturbing the bricks, they should be raked out to a depth of at least 15 mm, great care being taken not to damage the edges of the bricks. This will cause the joints to become much larger and spoil the appearance of the walling. After cleaning down the brickwork should be repointed as previously described. If the existing walling was built with soft facing bricks, the mortar to be used for the re-pointing must match the general work but in no case should it be stronger than 1:2:9 cement, lime and sand. Sulphate resisting cement may be used instead of Portland cement. For medium strength bricks the matching mortar must not be stronger than the existing

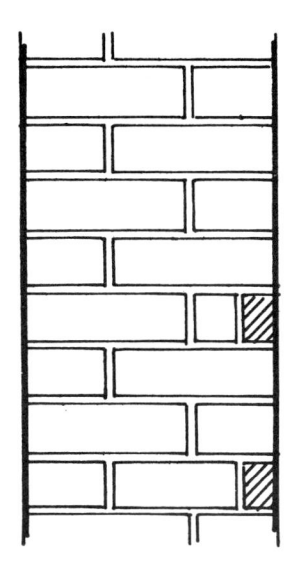

**Section
Brick slips built into wall shown cross hatched**

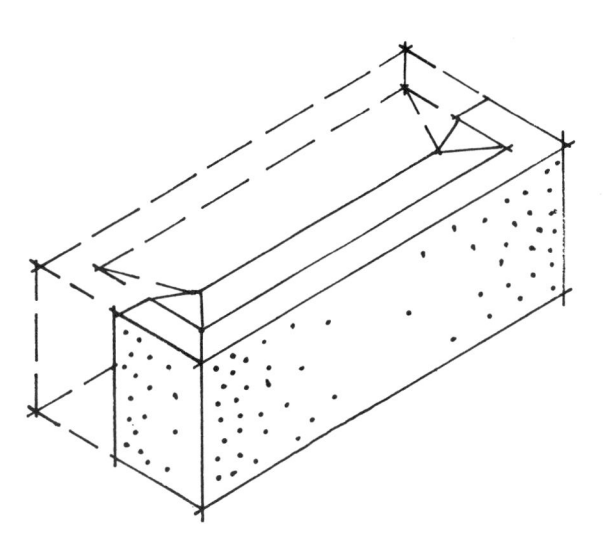

Brick slip 50 mm thick

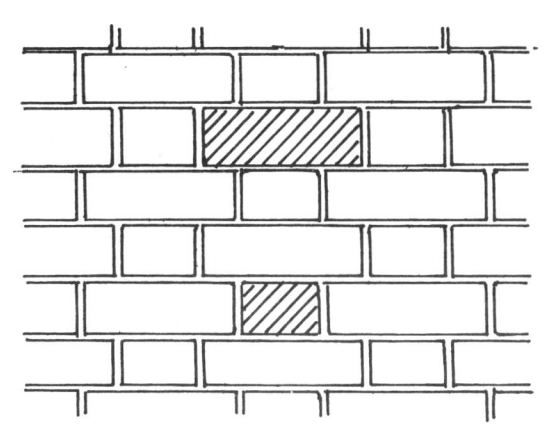

Elevation

Figure 45

bed joints and, as a general rule, not richer than a 1:1:6 mix.

Spalling of the bricks usually affects only a small proportion of the walling as a whole. If the general walling is sound then the damaged bricks can be cut out to the required depth as previously described. In some cases where the matching bricks are strictly limited it may be possible to salvage some bricks from existing walling and to cut some slip tiles from them on a mechanical saw. The thickness of these 'tiles' should not be less than 50 mm and these may be bedded into the spaces where the spalled bricks have been removed (Figure 45). This method will allow two brick faces to be obtained from each brick in the case of stretchers and, with care, it is possible to gain four header faces.

Matching such repairs with the general walling is very difficult and requires much care and patience.

Serious deterioration of chimneys
Chimneys are extremely susceptible to sulphate attack, not only from the sulphate content of the bricks, but also from the products of fuel combustion. For example, the three main products obtained from the burning of coal are water, sulphur dioxide and carbon dioxide. These dioxides, when mixed with water, will produce sulphurous acid from the sulphur dioxide, and carbonic acid from the carbon dioxide. Both of these acids may be harmful to the mortar joints and even to the bricks themselves.

In severe cases of sulphate and acid attack the chimney stack must be taken down to a safe level where the brickwork is sound. Usually this is below the damp-proof course level. The damp-proof course, apron flashing, back gutter, soakers and stepped flashing should all be carefully inspected for damage. Generally it is better to renew them, but, if they are in a sound condition, they may be re-used and the chimney rebuilt using a mortar comprised of sulphate resisting cement, lime and sand. Flue liners should also be built into each of the flues and these should be surrounded with a loose infill of vermiculite, expanded clay or any other similar effective insulating material. The chimney should also be finished off with chimney pots with flaunching around the top. Every corbel or oversailing course should be provided with a fillet on its upper surface to give a good weathering to the brickwork and as much protection against the weather and sulphate attack as possible.

If the chimney is old and has only been pargetted and not protected with flue liners, and if it is to be used in conjunction with a gas appliance, the rebuilding of the stack will provide a good opportunity to insert a flexible flue liner. This should be taken right down to the point where the flue of the appliance enters the main flue.

Cracking in the walling
There is quite a distinction in appearance between cracks that run more or less diagonally following vertical and horizontal joints alternately, and those that travel through alternate vertical joints and the intervening bricks and mortar beds. If these cracks permit the penetration of rain, all the cracked stretchers should be cut and the joints along which the cracks have passed raked out to a depth of at least 15 mm. When all the holes have been well brushed out and dampened they may be filled in with new bricks using a cement, lime and sand mix and all the mortar joints should be re-pointed with a matching mortar.

In the case of diagonal cracking the method to be adopted will depend upon the seriousness of the defect. If there has been movement due to frost heave, or expansion and contraction in the foundations, especially in clay soils (this movement may be accentuated by the close proximity of trees which may send their roots before the foundation level), then the cracking may be quite severe. In such cases the bricks on each side of the crack should be cut out, salvaging as many as possible. Where the cracking has just affected the joints then these may be raked out thoroughly. The brickwork should be rebuilt at each side of the cracked work where it was cut out ensuring that each brick is bedded solidly, particularly the underpinning. The rebuilt work should then be re-pointed. The mortar should be of similar strength to the mortar in the existing walling. Normally a mix of 1:2:9 would be quite adequate or, if a strong mix is required, a mix of 1:1:6, but the former is preferred. If the existing work was built with a weak mortar, this may be easily checked when raking out the joints, and in this case a mix of 1:3:12 would be strong enough. Overall, the use of very strong mortars should be avoided. A well graded sand in the mortar will also keep drying shrinkage to a minimum. It may also be an advantage in all such types of repairs to use a water retaining additive in the mortar.

If the amount of repair work to be carried out is fairly extensive then it may be advisable to build specimen panels as described in Chapter 3 before commencing repairs. The remedial work to be done before the repairs are started would depend upon the results of the initial survey.

If the cracking was due to movement in the foundations then the type of soil should be investigated. Trees planted adjacent to the building should be removed, as the suction of water from the soil by the roots will accentuate the soil movement. The further away the trees are from the building the less will be its effect. As a general guide, a tree's extend as far below ground as its branches do above. The depth of the foundations also plays an important part since move movement is found near the surface and those which are shallow are vulnerable to movement particularly when dug in clay soils. Usually if the trees are removed, the degree of movement will be greatly reduced, but where the movement cannot be easily controlled then it may be necessary to extend the foundations to a greater depth. This should be done in short lengths (see Chapter 7).

In the unusual case of frost heave, which may occur in shallow foundations built on a chalk sub-foundation, the deeper the level of the foundations, the less risk there is of frost affecting them. However, where there is danger of frost heave, extra protection can be given by laying an apron around the outside of the building and draining any water that falls on it away from the structure by means of gullies, or laying a land drain around the building. These will assist in keeping the sub-foundations reasonably dry and will prevent frost heave or keep it to an absolute minimum.

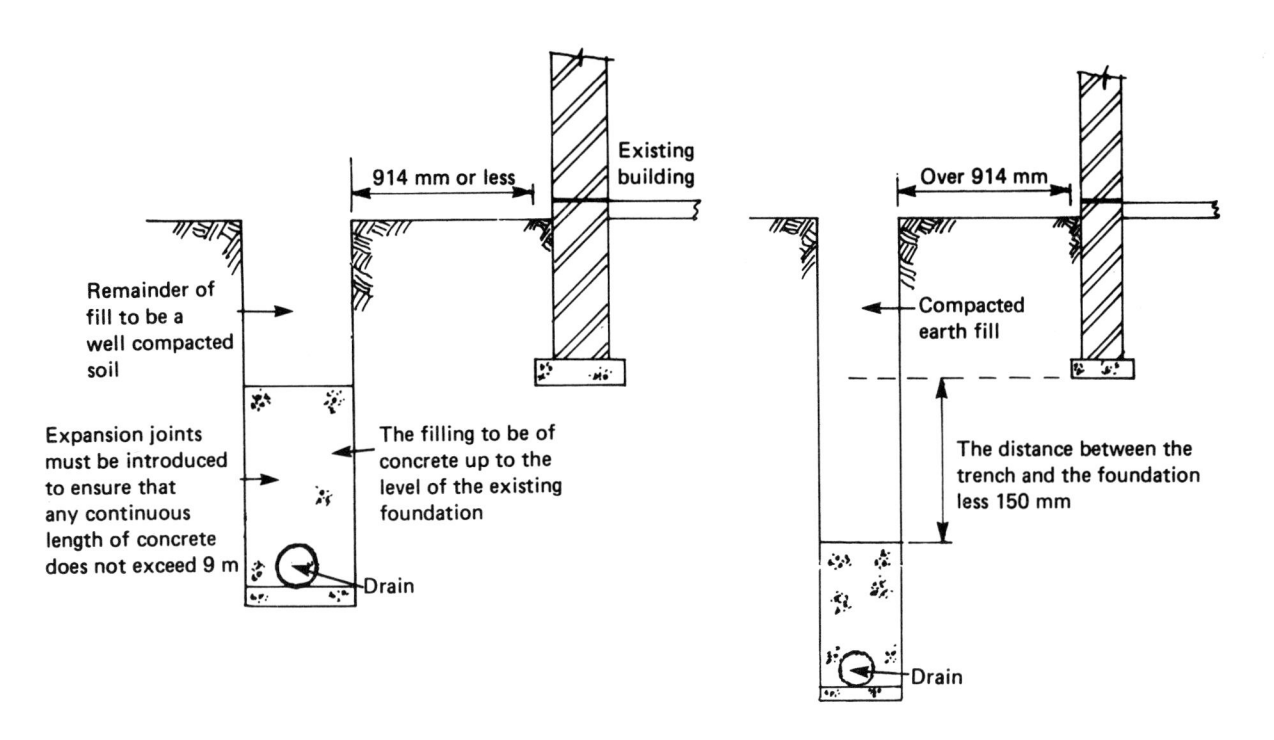

Filling in trenches which are adjacent to existing buildings

Figure 46

Mining subsidence can also create a real problem and any damage caused by this would require careful examination by a structural surveyor and each case carefully considered before carrying out any remedial work.

Where an excavation is made adjacent to a building movement may be caused in the foundations by undermining or even by the soil under the building drying out and shrinking. The building, therefore, must be adequately shored up to prevent any movement during excavating operations. On completion of the excavations, the trench should be filled with concrete up to the level of the underside of the existing foundations if the trench is within a metre of the building or up to a depth of the distance from the foundation less 150 mm (Figure 46).

Thermal movement caused by a wall being restrained at its ends may be overcome by cutting a chase about 15 mm wide on each face (preferably not in line but staggered), and filling the gap with a non-hardening compound or mastic, with the face of the joint being covered with a metal strip clipped into the chase.

The only remedy for vibration caused by heavy traffic is to divert the heavy vehicles. This is a serious problem which has been inflicted on many towns and villages.

Corrosion of iron and steel

Where the defects are due to corrosion it will be necessary to open up the brickwork, salvaging as many bricks as possible, to gain access to the offending steelwork. If the corrosion is not too far advanced the metal may be thoroughly cleaned and primed with a rust inhibiting paint and coated with bitumen before rebuilding the brickwork.

If the corrosion is well advanced then it will be necessary to replace the steelwork. This would subsequently become a major repair exercise and sufficient brickwork must be removed to allow access for the removal of the defective steelwork and for its replacement.

When the corroded steelwork has been replaced the wall may be rebuilt and all practical steps must be taken to prevent moisture penetration by the introduction of suitable damp-proof courses, and by ensuring that where the steelwork enters the brickwork it is coated with a bituminous compound and also that the wall is well caulked at the point of entry.

If there has been a considerable amount of demolition work, due to the extensive corrosion, necessitating complete reinstatement of the steelwork, excessive weakening of the wall could result. In such cases the rebuilding work may be strengthened by using reinforcement in the bed joints. Such reinforcement may be of the expanded metal type or welded fabric, both of which are obtainable in wall sizes and are very convenient to handle. This method of reinforcing brickwork is most advisable particularly if the wall is to carry any imposed loading.

Unstable cavity walling
Wall ties which have corroded because the galvanising or other protection has failed will continue to rust and expand until no metal remains. This may well result in an unstable condition in the cavity wall. The defects will be particularly noticeable at door and window openings.

The inspection of the extent of the corrosion in the wall ties may be carried out by cutting out the bricks overlying the ties. This will also indicate the depth of the corrosion within the outer leaf. Another method is to remove bricks at the corners and to view the wall ties through an endoscope (a fibre optic device for remote inspection). Unfortunately, however, this method only allows inspection of the ties within the cavity, does not indicate the extent of any corrosion that may have taken place inside the wall, and may not reveal a loss of connection between the leaves.

Locating the corroded ties will be comparatively simple if staining has occurred on the face of the wall or if the joint has been affected by expansion, but in not so obvious cases, a metal detector may be used. It is, however, advisable to have a demonstration on the wall before purchasing one as there is a variety of such detectors on the market – some will only detect ferromagnetic materials. Therefore, if other metals are to be detected, the supplier should be consulted. Also the thickness of the wall may have an effect on the results obtained.

The weight and shape of the instrument should also be carefully considered as it will have to be used on vertical surfaces and carried up ladders and used on scaffolds. It should be waterproof and have its own power supply.

The method of repairing a defective cavity wall will depend upon the extent of the corrosion and damage. Where the wall is not too seriously damaged the bricks overlying the ties may be removed. If these bricks are to be salvaged a portable dry carborundum disc saw may be used to cut out the joints, but it is important that protective clothing and safety goggles are worn, particularly as there is a great possibility of the disc striking the metal tie in the wall.

The ties may then be removed fairly easily from the inner leaf by loosening the mortar on each side of the tie body to free the fishtailed splines. Power chisels or masonry drills may also be an advantage with this operation. There are also power devices commercially available which will grip the tie and extract it from the inner leaf. In the case of wire ties they may be cropped with a pair of cutters or they may be bent back flush to the inner leaf, but care must be taken to ensure that they do not interfere with the operation of inserting the new ties.

Preferably the new ties should be of stainless steel and may be of the shape shown in Figure 47. The replacement ties are inserted in the cavities left by the removal of the corroded ties. If these are unsuitable, fresh holes may be drilled into the inner leaf with the aid of masonry drills and the tie inserted and resin grouted in the inner leaf, and re-bedded in the outer leaf when the bricks are replaced. (Resin grout should not be used on water saturated walls.) A resin injection gun is used to grout the ties. In saturated walls grouts comprised of mixtures of Portland cement fillers and water based polymers may be more suitable and these may even perform better than resin grouts in damp walls.

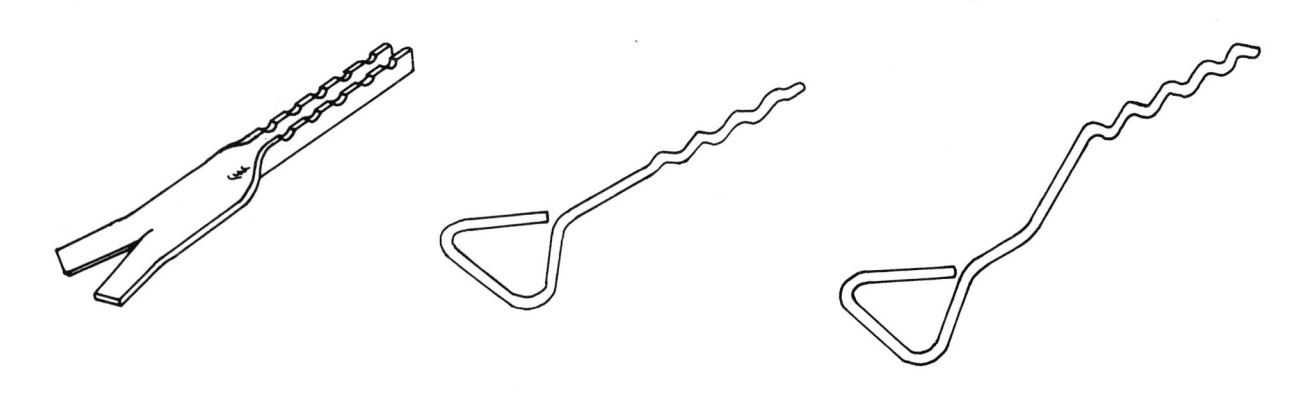

Types of wall ties used for the repair of cavity walling
Figure 47

In severe cases the outer leaf may also be affected by sulphate attack or frost action, in which case it should be demolished, ensuring first of all that all the roof and floor loads are carried securely by the inner leaf. If they are not suitable props must be provided and the new ties may be inserted in the inner leaf and the outer leaf re-built.

In instances where the ties have become badly corroded but the leaves are in a good condition, specialist methods may be used to stabilise the two walls. Such methods include the use of resin grouted ties, also by inserting expansion grip fixings. The resin grouted ties can usually be installed using standard tools such as a masonry drill, a driving attachment, an alignment aid, a capsule insertion guide and a resin injection gun. Basically, a hole is drilled through the outer leaf, preferably not through a perpend as these are not usually completely filled, and then partly into the inner leaf. The specialised tie is inserted into the hole and sealed by either grouting the inner leaf with a resin capsule and gun grouting the outer leaf, or by pumping resin through the hollow tie via a grommet. This fills the hole in the inner leaf then flows back over the tie to fill the hole in the outer leaf.

Installation of expansion grip fixings requires only a masonry drill and a controlled torque or pressure device. The holes are drilled through the outer leaf and part of the way into the inner leaf and the expansion grip fixing is inserted and tightened by means of a screw cone or screw/expander plug up to 3 to 7 Nm.

Damaged corbels etc.
Oversailing courses may be badly damaged by erosion or settlement but once again it must be emphasised that remedial work be carried out before repair commences. Where the bricks have been badly eroded and the remedial work completed then the shapes of any moulded bricks that may be incorporated in the corbels, sills, string courses etc. must be carefully detailed. The replacement bricks should be cut to the required shape (usually by hand), or replacement bricks ordered from the suppliers if the mouldings are standard units or, if sufficient numbers are required, templates produced and sent to the brick manufacturers as described in Chapter 1.

When the replacement bricks are ready to be installed the defective work should be carefully cut out, using a minimum amount of vibration, and the cavities cleaned out and dampened. The new bricks may then be built in ensuring that they are well bedded and underpinned, especially where the replacement work is on oversailing courses. If the rebuilt work is for corbels or sills, which are overhanging, the bricks must be lined up on their underside and not at the top of the course, as the lower edge will be the 'eye line'.

If there is any exposed surface at the upper side of the corbel, string course, or sill, this should be provided with a weathering either in the form of a fillet, using a mixture of sulphate resisting cement and sand, or by laying the top course to a slope, or a sloping surface incorporated in the moulded bricks. Soft bricks are not recommended for work of this nature because of their vulnerability to severe weather conditions. Therefore, hard burnt bricks should be used and bedded in a fairly dense mortar comprised of a cement:lime:sand mix in the proportion of 1:1:6. The cement, for preference, should be sulphate resisting to avoid the possibility of sulphate attack.

Arches which have been displaced should be carefully dismantled and the bricks, if unbroken, cleaned and numbered from each springing point, 1L 1R, 2L 2R, 3L 3R, and so on to the key brick, and stored on one side. One or two courses above the arch should also be removed and the whole of the opening thoroughly cleaned. A suitable arch centre is then made and fixed in position in the opening. The skewbacks are then checked for accuracy and soundness and rebuilt if necessary. The positions of the arch voussoirs are then marked around the arch centre and nails fixed at striking points. A length of line is attached to each nail to provide a means of checking that the voussoirs are normal to the curve.

The voussoirs are then built in from each side of the arch until the key brick is placed in position. The whole work is carefully checked for straightness along its face and then the joggles are grouted in. The arch should then be allowed to set before the brickwork over the top is rebuilt. This should be done solidly and the underpinning filled in soundly.

Deterioration in parapet walls
Parapet and copings are subjected to severe and extreme weather conditions and are, therefore, very liable to sulphate attack. Copings are prone to becoming detached from the walling and the joggle joints broken or the cramps seriously corroded. In severe cases this can create a serious hazard as the copings may become completely dislodged by a strong wind. The copings should be removed from the wall and the joggles cleaned out and the stones checked for soundness. If the cramps have been used then these should be checked for any serious corrosion. If they are reasonably sound then they should be cleaned ready for re-use, or, if badly corroded, they should be replaced. If the parapet wall is badly deteriorated due to sulphate attack it should be dismantled at least down to the lower damp-proof course which should be carefully checked for soundness. If there is any doubt whatsoever regarding the damp-proof course then it makes economic sense to replace it. The parapet wall should then be rebuilt with hard burnt

bricks and a fairly dense mortar comprised of a mixture of sulphate resisting cement, hydrated lime and sand. A new damp-proof course should be installed just below the coping. It is unwise to re-use the old damp-proof course which may have been in the old brickwork and which would be removed when the walling was demolished.

The coping stones or brick coping, as the case may be, should then be replaced and the cramps or joggle joints grouted.

If the interior face of the parapet wall had been rendered this may well have accelerated the sulphate attack. When rebuilding, therefore, it would be wise to forego the rendering and use fair faced brickwork.

Dampness in buildings

Because of high remedial costs it is essential that a positive diagnosis is made to distinguish between rising damp and other sources of dampness. The presence of damp that has existed for some time is indicated by clearly visible signs such as damp patches on walls, peeling and blistering of wall decorations, patches of efflorescence and mould growth. The presence of rising damp, however, is often indicated by a horizontal tide mark. There may also be a concentration of hygroscopic salts which have been absorbed from the soil, bricks and mortar and these may also cause dampness in humid or wet weather.

The cause of rising damp may be due to:

1   A damp-proof course becoming damaged through uneven settlement of the building.
2   The damp-proof course becoming punctured or damaged through the weight of the building.
3   The bridging of the damp-proof course particularly at the base of cavity walls, or earth or other materials being placed against the external face of a solid wall above the damp-proof course.

Where rising damp is positively diagnosed, the first step is to ensure that the damp-proof course is not bridged. Externally, any material that may be piled above the damp-proof course should be removed. If the dampness is in a cavity wall, corner bricks should be removed so that the cavity may be viewed for any blockage. If there is a blockage a length of pipe which has been flattened at one end and bent over at 90 degrees may be inserted into the cavity and any loose material raked towards the corner and removed. If the blockage is hard and not easily removable by the length of pipe, then cleaning holes should be cut at intervals along the wall and the offending debris removed. During this operation the ventilators should also be carefully inspected and cleaned out to ensure that any underfloor space is well ventilated.

Should there be no sign of bridging of the damp-proof course and there is no indication of water penetration from another source, such as a defective plumbing system, it may be assumed that the damp-proof course itself is defective, in which case it must be replaced – an expensive and slow task. Hence the importance of checking every possible cause of the defect before taking this step. In cavity walls or walls of one-brick thickness built in lime mortar, or a fairly weak cement:lime:sand mortar, a horizontal joint can be cut in about 1 metre lengths with the aid of a tungsten-carbide tipped chainsaw. When a length is cut a membrane is loaded with mortar and inserted, with temporary wedges driven to prevent settlement. The mortar is forced into the cut to ensure that it is solid. This is repeated right along the defective wall, the membrane being well lapped at each section. Before starting this work it is most important to check that all electrical wiring, gas and water pipes are moved out of the way.

If the floor adjacent to the defective wall is a hollow timber construction, the new damp-proof membrane must be below the lowest member of the timber floor, usually the wall plate. If it is a solid floor then it must be situated near to the upper surface, and in this case the joint between the solid floor membrane and the new damp-proof course which is being inserted must be closed by extending the membrane to form a vertical damp-proof course between the solid floor and the horizontal damp-proof course, or any other form of tanking type of treatment.

A new damp-proof course can be installed in a thicker wall by cutting out a complete course of bricks which would have to be done in about 1 metre lengths, then inserting the new damp-proof course and rebuilding immediately, ensuring that the wall is solidly underpinned. The work should then be allowed to set before the adjacent section is commenced, although work in another section of the wall may be started in a manner similar to that described for underpinning in Chapter 7. This method is, of necessity, slow and, of course, expensive.

A more economical method in this case would be to mask the effects of the rising damp by using a wall lining technique, such as fixing plasterboard on timber battens. These timber battens must be pressure impregnated and, where they are sawn, the ends painted with a preservative to ensure full protection against attack by rot. Alternatively, a proprietary system can be used for the same purpose. This, of course, does not cure the rising damp problem but does allow the interior of the building to be reasonably dry. The use of a dry lining, however, may cause long-term problems as the damp may increase its height and intensity through the reduction of evaporation and may pose dangers to wooden components such as window frames. However, where alloy frames are installed this problem can be overcome.

Another method of overcoming rising damp in thick walls may be used by chemical injection. These are proprietary methods and may involve the use of silicone or aluminium stearate water repellants. These may be injected under pressure or transfused into closely spaced holes drilled in the brick or mortar courses along the damp-proof course line. Which ever system is adopted care must be taken to ensure penetration through the entire thickness of the wall. These methods generally give the best results in the summer or late summer months when the walls are at their driest and the rising damp at its lowest.

In all cases of rising damp, any timber construction, such as floors, frames etc., must be carefully checked for the presence of wet or dry rot. Where mould growth has occurred this may be cleaned off with a fungicide as described in Chapter 4, but interior walls do not usually require more than one application to eradicate the growth.

Damp walls may also be rendered with a rich cement: sand mortar undercoat in the proportion of 1:3 (with possibly a waterproof additive or using a waterproof cement), and finished with a class C gypsum plaster. This will give a dry interior but may present problems with condensation, so good ventilation must be provided to offset this.

Other defects permitting the penetration of water are:

1. Defective damp-proof courses in parapet walls. The treatment for this is the same as described under the section 'Parapet walls' and replacing the damp-proof course when the wall has been taken down.
2. Faulty flat roofs. This is specialist work and generally necessitates resurfacing as patching rarely works efficiently.
3. Dirty wall ties. The treatment for this is the same as that for the removal of corroded wall ties.
4. The covering of a steel lintel or damp-proof course on the lintel with a rendering. This is a defect which has occurred in modern structures in which openings have been bridged with steel lintels and the outside walling rendered or surfaced with a waterproof proprietary coating. If this rendering is taken right over the steel lintel it may allow the penetration of water between the lintel and the frame (Figure 48). This problem may be overcome by cutting the joints and the rendering away from the frame and pointing the joint between the frame and the steel lintel with a non-hardening mastic.
5. Cavity insulation materials which have been badly installed. This is an extremely difficult defect to overcome and usually means cutting out corner bricks and raking out the defective material or cutting out the bricks at the damp patches, then cleaning out the cavity and rebuilding.

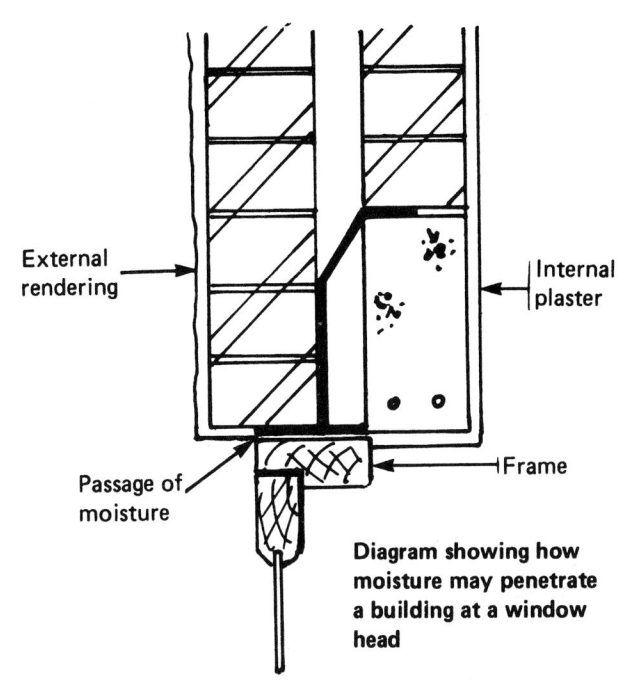

Figure 48

Diagram showing how moisture may penetrate a building at a window head

Bulging walls

It is not always easy to judge from appearance how far damage to brickwork is likely to affect the strength or stability of the structure. No firm rule can be applied but a check should be made on the bearing area of joists, lintels and roofs to ensure that there has not been a significant loss of support. Any flat arches or brickwork over openings should not show wide cracking at the abutments or skewbacks.

If a solid wall shows a bulge on one face which is not visible on the other then it should be treated with extreme caution. Similarly, with cavity walls the effects of bulging or leaning can be very serious and needs careful inspection.

When considering what repair treatment may be necessary for such defective walls the following factors should be taken into account:

1. The position and bonding of cross walls, buttressing walls and piers which provide the wall with lateral restraint.
2. To ascertain if the upper floors, roofs and walls are liable to exert side thrust on the defective walling.
3. The nature and condition of the wall and its foundations.

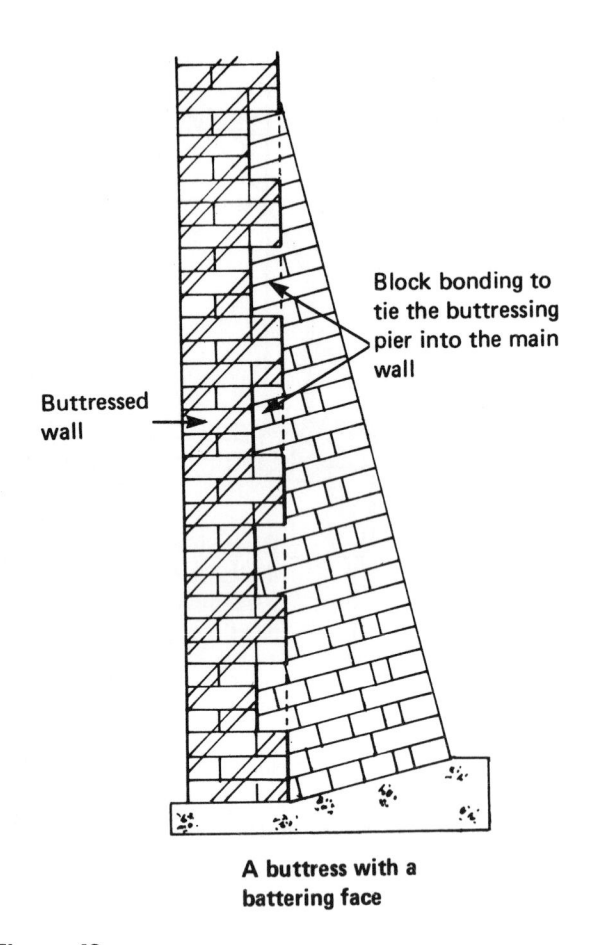

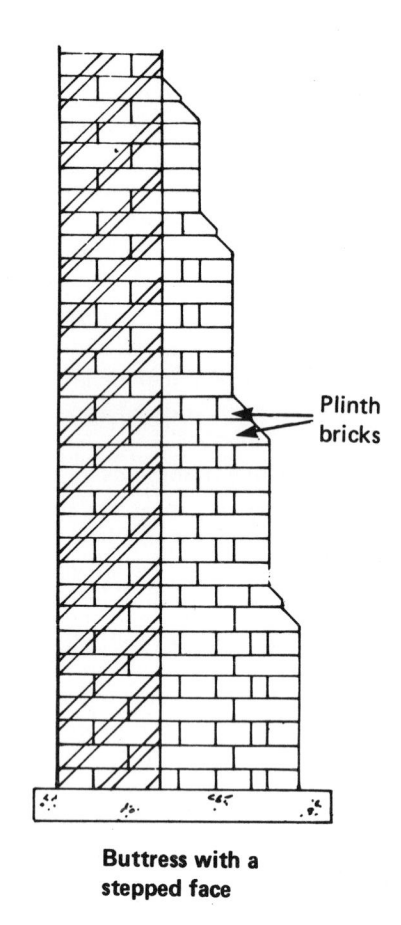

Block bonding to tie the buttressing pier into the main wall

Buttressed wall

A buttress with a battering face

Figure 49

Plinth bricks

Buttress with a stepped face

Figure 50

4   If there is any possibility of vibration being caused by nearby traffic or even by the user of the building.

Once the survey of possible causes of the damage is completed, steps must be taken to correct the faults or prevent further movement or bulging. If, in spite of the bulging, the walls are quite sound and the imposed loads are being carried quite safely, to prevent any further movement tie rods may be passed through the structure in the thickness of the floor or at roof level. These should be anchored to another wall, either external or internal, that is sound or may even be tending to move in the opposite direction. The rods may be threaded at the ends and tightened up with nuts having first placed steel restrainers on the rods. This method is quite common and one which is in universal use throughout the UK, and will give many years of good service.

Another method that may be used is to build buttressing piers against the defective wall. These should be placed, if possible, immediately opposite the direction of any thrust that may be acting upon the wall. The pier may have a sloping face towards the defective wall and if it is built off a sloping foundation which is parallel to the bed of the pier there will be less possibility of 'slip' occurring (Figure 49). The foundation of the pier must be of sound construction and capable of resisting any thrust that may be imposed upon it. This is most important because if the pier is not carried safely, any slip in the foundation could increase the thrust upon the defective wall and add to any settlement which has already taken place. Buttressing piers may also be built with a vertical face but reducing the length of the pier as it gets nearer to the top. The top of the pier may be finished off with 'tumbling-in' courses to provide a suitable weathering (Figure 50).

Building a vertical pier makes it easier to bond it into

the defective wall though the common practice is to block bond the work, usually every alternate three courses. Similarly the sloping faced piers are also block bonded, but usually these require much more cutting out of the wall to allow adequate tying in of the new work. In such cases the cutting out of the indents must be done with care and as little vibration as possible to avoid creating any further damage to the defective wall. A portable masonry cutting disc would prove very useful for this work rather than using a hammer and chisel. The new brickwork should be matched with the existing work as previously described.

If the situation is such that neither of the two methods described can be used with confidence then the defective wall should be pulled down and rebuilt. Care, however, must be taken to ensure that all floors, roofs and any imposed loads are adequately supported with props.

The mortar mix recommended for the rebuilding and the buttressing piers would be a Portland cement:hydrated lime:sand in the proportion of 1:1:6. It is not advisable to use a stronger mix as it may cause cracking between the old and the new work. If a weaker mix is used it may give insufficient resistance to the movement in the wall and, additionally, may not stand up to the weather conditions.

Flood damage
Before starting any repair work on a structure that has been flooded it is most important to test the services, particularly electricity and gas, and isolate if necessary. Check for any water trapped in ducts, cavities, pits and underfloor areas. This water should be pumped out or holes cut or drilled to drain the water away. Clear any mud and debris away from ventilators and from external walls where it may have piled up above the horizontal damp-proof course and from the cavities. This may be raked out by removing corner bricks or occasional bricks along the face of the wall. If the mud is too difficult to remove by raking it may be flushed out with a hose pipe, but extreme care must be taken to ensure that no further damage is inflicted upon the structure. After cleaning, the structure should be dried out with the use of heaters. Portable paraffin heaters are useful for this purpose but these produce considerable water vapour. Therefore, the building must be well ventilated and doors and windows should be kept open and floorboards lifted to increase the draught.

If any salts appear on the surface of the wall these should be brushed off. This is best done when they are dry. The appearance of these salts indicate that the wall is drying out.

Brickwork, in general, will withstand flooding very well but erosion may occur to soft brickwork or sand lime bricks especially where the flooding has been due to sea water. But this is usually a long term effect and if this happens then the repair work must be carried out in the same manner as that described for damage through frost action.

The most likely damage to occur in the case of flooding is to the foundations, especially where they are situated in shrinkable clays. These swell when wetted and are likely to cause movement in the building which may result in severe cracking in the brickwork. In these cases the ground should be allowed to dry out before deciding upon any remedial action to be taken. There may also be scouring out or erosion from the foundations or even the base of the wall, in which case this must be carefully inspected and remedial measures taken. This may necessitate the reinstatement of the foundations (see Chapter 7) or demolishing the damaged walling and rebuilding.

Drains should be checked and rodded to clear away any mud. In difficult cases it may be necessary to use mechanical means to clear silt which may block up long lengths of drain. As in all cases of blocked drains it is advisable to rod against the flow of the drain rather than down the gradient, since the latter method tends to compact any blockage, making it more difficult to clear.

Subsidence
Subsidence in foundations is dealt with in Chapter 7, but there is a further aspect of subsidence that creates a serious problem. This often occurs on sloping sites where the ground floors have been built out of the ground or where foundations have been dug deeply and the infill under the solid floor has been rather excessive (it is now recommended that this should not exceed 600 mm). If this is not adequately consolidated then it will gradually sink and consequently the ground floor will become deformed and cracking will occur where the weight of the partitions cause it to distort, also around the edges of the floor against the external walls and any other points where there may be loading. This uneven distortion will inevitably be transmitted through the partition walling.

The external walling may be quite sound and the partition walls well bonded into them at the junctions so any subsidence which takes place along the length of the partition will create stresses resulting in serious cracking. This may also have the effect of distorting door frames and causing doors to jam, and there may also be distortion in the staircase, all of which adds up to a very serious problem which is very expensive to repair.

To carry out the remedial work the building should first be evacuated and all furniture removed, the partitions demolished, first propping up the upper floors if necessary, then the ground floor lifted. The infill should then be inspected and either stabilised by pumping in grout, or removing and replacing with a weak concrete or a hoggin, but consolidating soundly in thin layers. Another method is to remove the infill and lay a site concrete at the base of the excavation, building up sleeper walls and installing a hollow timber floor in the place of the original solid floor. The concrete should be thickened where the partition walls are to be erected. The partition walls may then be rebuilt and door frames renewed and doors rehung. Distorted staircases should then be carefully examined and if it is possible to correct the distortion then the wedges should be removed, reglued and replaced. Also check that the newels are soundly supported.

If a solid floor is required to be relaid then a damp-proof membrane must be inserted and closed with the horizontal damp-proof course in the external walling.

If the partition walls are load-bearing and are required to carry the upper floors, the concrete at their base must be thickened so that it will carry the weight of the partition, plus any imposed loading, safely. The partitions must also be well bonded into the main walling to obtain maximum stability in the structure. Extreme care must also be taken when cutting and pinning up to ceilings, soffits of concrete floors, wall plates etc. to ensure that the joints are solid. Any pinning up must be allowed to set and harden for at least 7 days before removing any props supporting upper floors etc.

If the partitions are non-load-bearing then it would be advisable to consider the use of dry linings as their replacement in order to reduce the weight on the foundation and the effects of drying shrinkage, and more importantly, to reduce the amount of water used in the remedial work and subsequent drying out time.

Good ventilation should be provided throughout the whole operation of the remedial work to assist in the drying out of the structure. This will allow the internal finishings to be installed and redecorating put in hand as soon as possible. If heaters are used to assist in the drying out process, good ventilation is essential to avoid a concentration of water vapour in the building. On the other hand too rapid drying out must be avoided as this can create other problems with drying shrinkage and so on. Dehumidifiers may be very helpful and should be used in conjunction with heaters especially during cold weather.

Although this section has dealt primarily with subsidence due to poor consolidation of underfloor fill there is a similar defect which may occur through the infill heaving or swelling, and in severe cases lifting the floor and even disrupting the walling surrounding the floor. The damage to partitions, doors and staircases etc. will be similar to that caused by subsidence and the remedial work very much the same.

This heave will occur when a shale is used as a hardcore. Under a solid floor red shale is very vulnerable, as when it becomes wet it will expand and cause the flooring to lift.

Fire damage

The effects of fire vary according to the severity of the fire, also the thickness of the walling and the types of bricks in the walls.

Where there has been spalling of the bricks due to sudden quenching after being hot, these may be cut out in the same manner as described for damage due to frost action. Similarly, where joints have been damaged they may be raked out and repointed.

Walls built with concrete or sand lime bricks may be so seriously damaged in the case of a severe fire that the structural stability may be affected and in such cases demolition of the walling and rebuilding may be necessary. More likely, however, is the distortion that may be caused by thermal movement of structural steel members built into a wall. These are likely to cause severe cracking, bulging or displacement of the walling.

Each defect must be closely examined and if the wall has bulged to a major extent or if there is severe cracking then it may well be a sound policy to demolish the wall and rebuild. If, however, the wall has only minor bulging and the cracking is fairly localised and repairable, remedial work may be done by inserting tie rods or building buttressing walls or piers to stabilise the wall and the cracks raked out and repointed.

Summary

With all forms of damage that may occur to brickwork, in order to ensure that repairs will be effective, it is necessary:

1   To know what caused the damage to the walling (diagnosis is not always easy).
2   That remedial work to eradicate the cause of the defect must be carried out before any repairs are put in hand.
3   That the new work must match the existing work as closely as possible.

All repair and renovation work is expensive and the urge to accept a cheap, short-term solution instead of the most efficient, should be avoided. However, it is also sensible to compare the costs of equally efficient methods and adopt the cheapest which will satisfy all requirements.

6 Repairing and restoring brick features

In Chapter 5 the repair of facework and general walling was described in detail. In this chapter the emphasis will be on the repair and restoration of special features in brickwork.

Before any work of this nature is started it is necessary to measure the feature in close detail, reproduce it in a drawing and then record the details in a specification. Such details that should be noted will include the size and shape of the feature; any special mouldings and moulded bricks involved in the work; and those which need to be cut or ordered from the manufacturers; the type of bricks used in its construction; the bedding materials and their probable mix and the type of pointing and the finish of the joints.

The drawings should clearly show the feature in elevation together with plans and any sections that may be necessary to identify the feature after it has been removed. Any numbering sequences of the bricks or units which are required to be salvaged should be clearly shown and the same numbers marked on the units as they are taken down, cleaned and carefully stacked ready for re-use. The drawing should also show any subsidiary work that may be required in conjunction with the restoration, together with its specification.

If the work necessitates the use of mouldings then these should be drawn to a large scale, or even full size, so that the replacement units can be reproduced accurately.

It is important to bear in mind that once the feature has been pulled down the reproduction will depend entirely upon the record that has been made, and any mistakes or omissions will be difficult to rectify.

Special features in brickwork may be produced in either fine axed work or gauged or rubbed work, and would include balustrades, cappings to parapet walling and piers, cornices, string courses, dogtoothing, block bonded quoins, plinths, blind panels, herring-bone and basket weave panels, brick lintels, arches of all shapes and niches. Figures 51 to 57 show examples.

Fine axed work

In this method the bricks are cut by means of the usual cutting tools, such as a club hammer and bolster, scutch, brick or comb hammer, and carborundum stone. Alternatively, the bricks may be cut by means of a

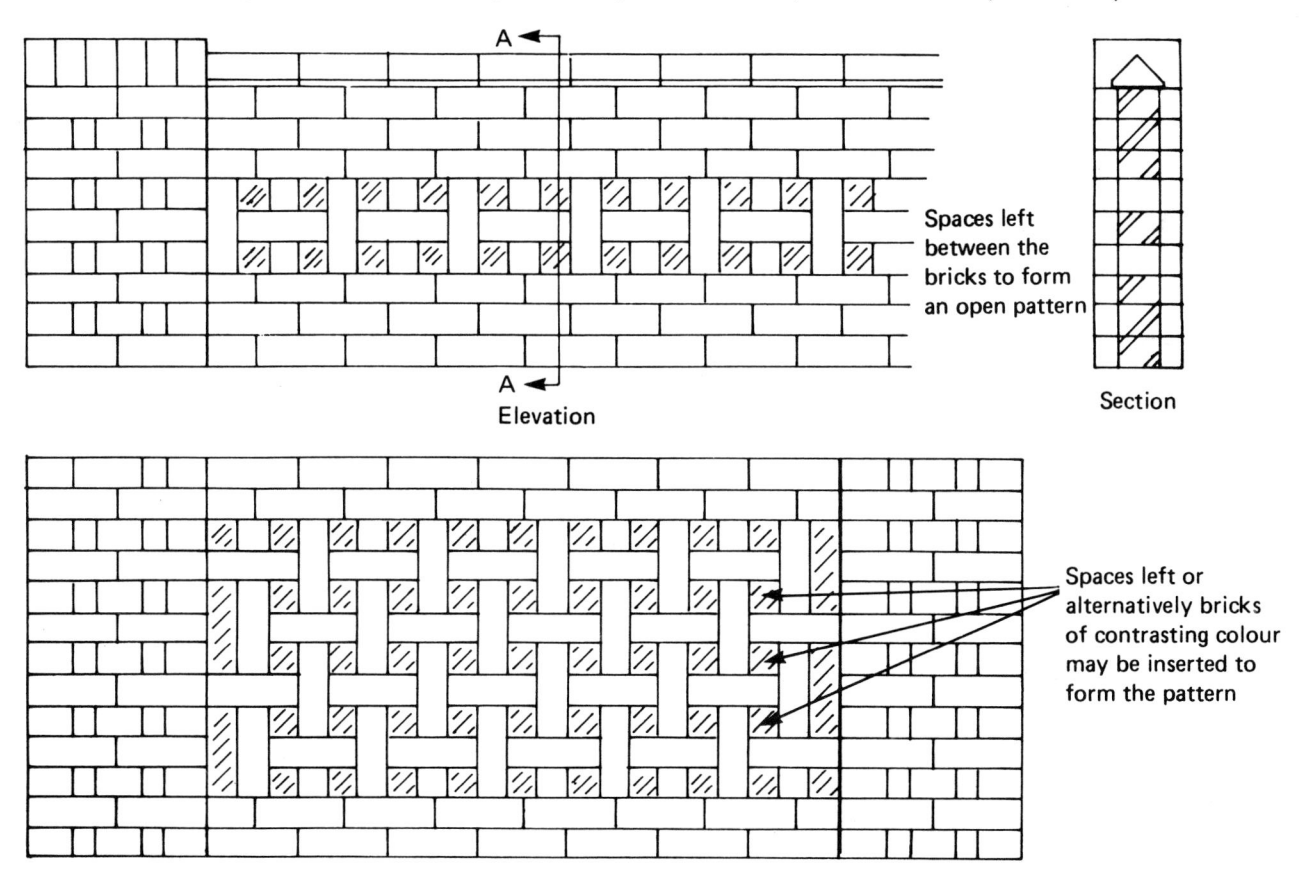

Decorative balustrade walling

Figure 51

Brick on edge and creasing tiles

Saddle back capping using purpose-made bricks

Feather edged stone coping

Nibs

Brick on edge and roofing tiles

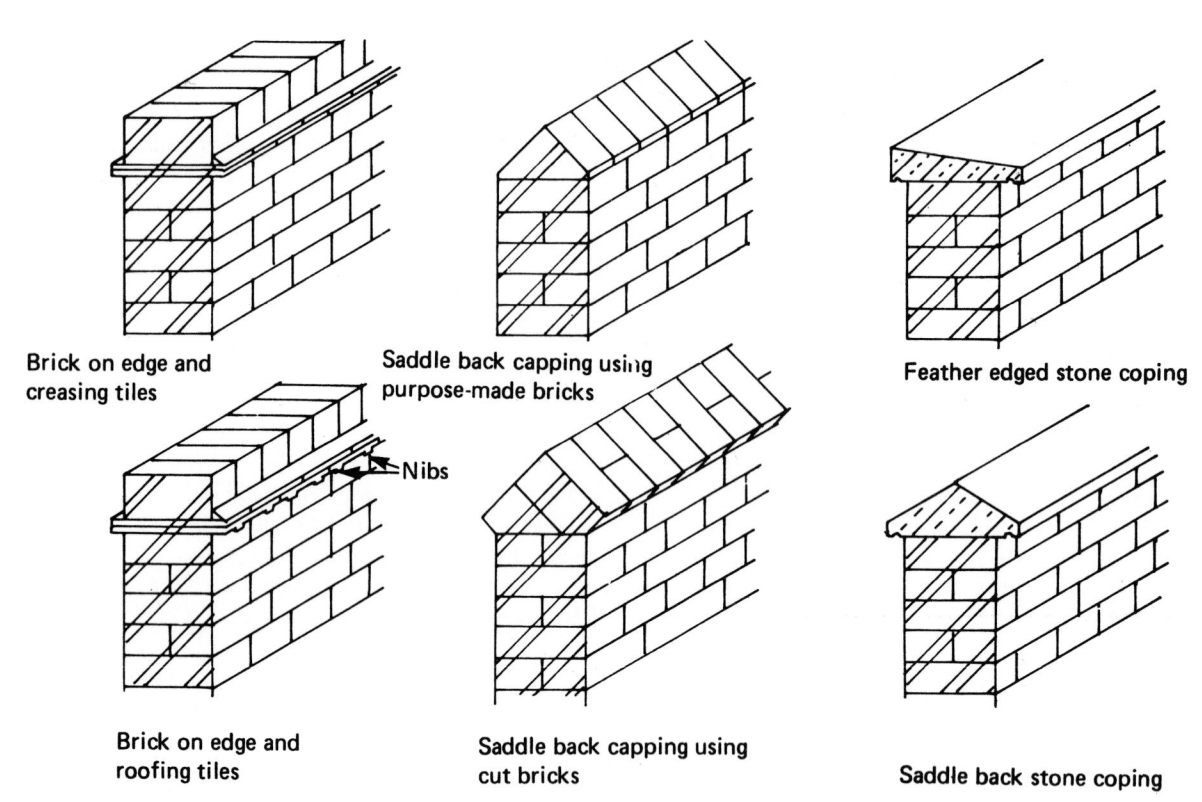

Saddle back capping using cut bricks

Saddle back stone coping

Typical cappings and copings to parapet walls

Figure 52

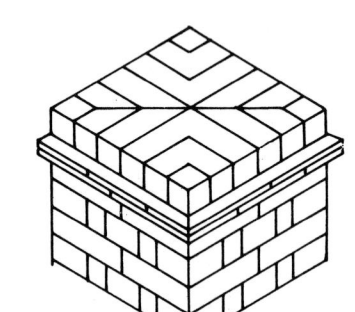

A mitred brick on edge and creasing tile capping

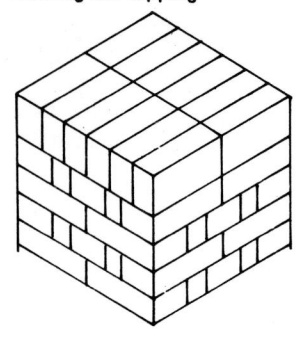

A simple brick on edge capping

Figure 53

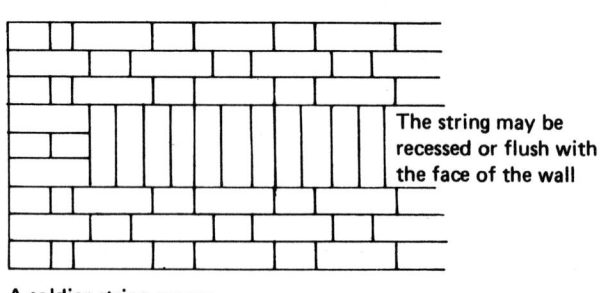

The string may be recessed or flush with the face of the wall

A soldier string course

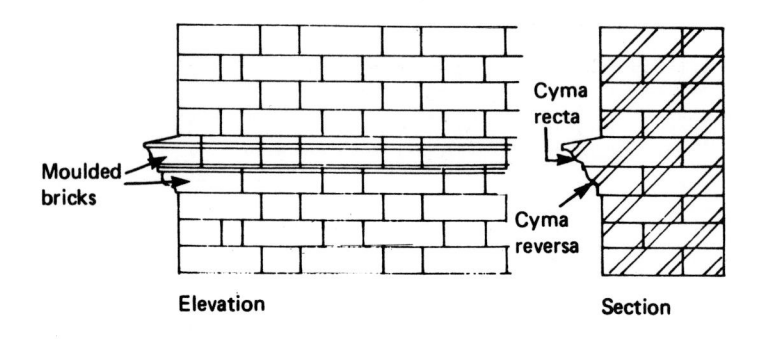

Moulded bricks

Cyma recta

Cyma reversa

Elevation

Section

Typical string courses

Figure 54

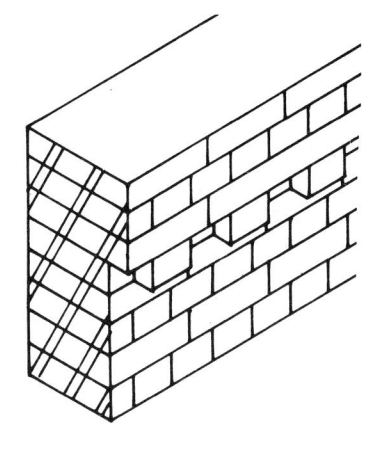

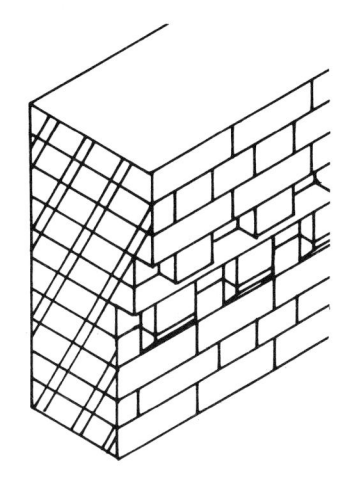

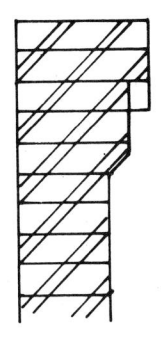

Section showing the lower dentil course supported by inverted plinth bricks

Dentil courses

Figure 55

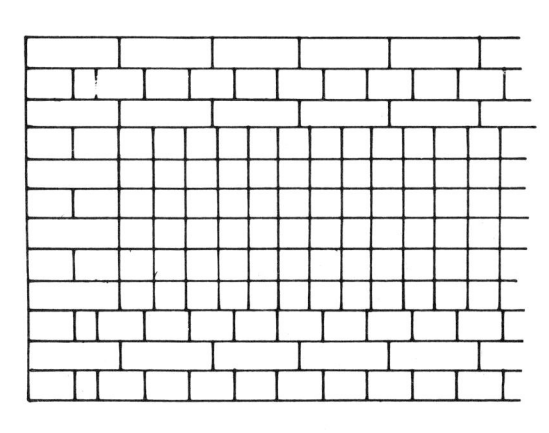

Elevation

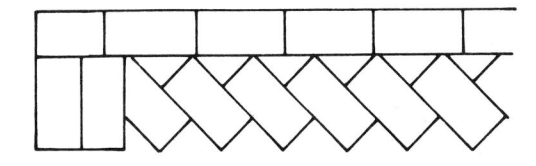

The recesses are usually plaunched with a portland cement and sand mortar to provide a weathering

The courses in the dog toothings are laid in opposite directions to form the bond

Plan

Dog toothing

Figure 56

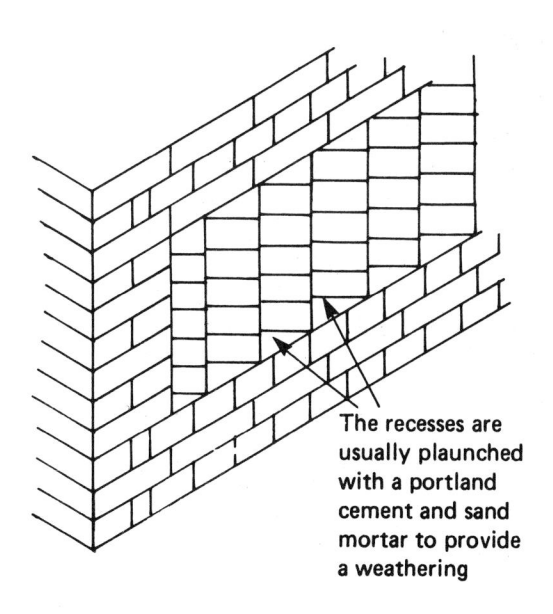

Arch shapes and their setting out

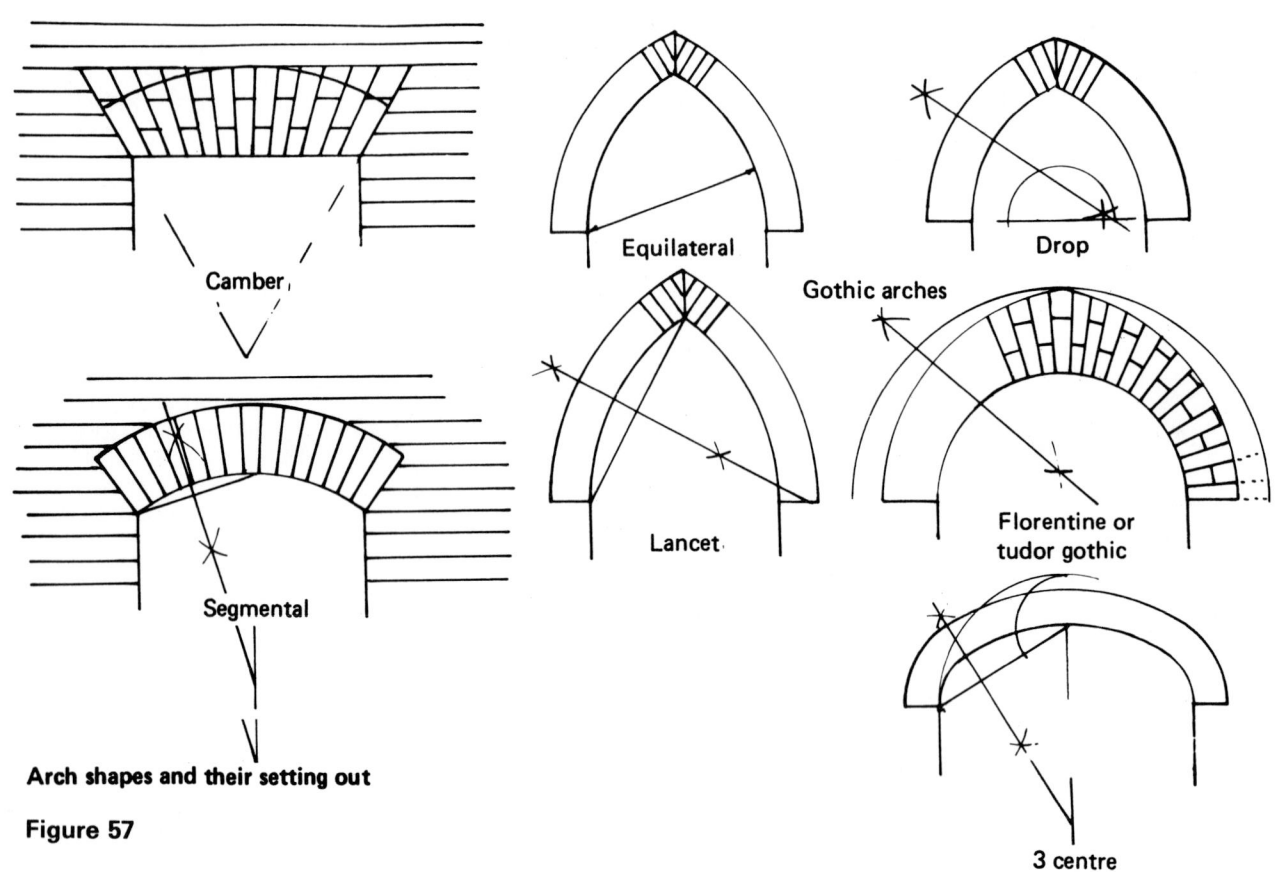

Figure 57

mechanical disc, of which there are portable and bench types on the market. Usually the best results are obtained from the bench type as the machine is more stable and the abrasive disc is of a larger diameter and can thus cut through the thickness of the brick at one pass. The blades should be of a type that do not shatter if they should become broken during operating. In any case protective clothing and safety goggles should be worn during the work.

Preparing the voussoirs for an arch

The following is an outline of the general procedure for preparing the voussoirs for a fine axed arch. This procedure may be varied slightly to suit the different shape of arches but the basic principles remain the same.

1. An outline of the arch should be set out full size on a setting-out board.
2. Mark out the positions of the voussoirs on the extrados of the arch. If the arch has a key brick then the voussoirs are set out by marking out the key brick first and dividing the extrados into a number of equal divisions which must not be greater than the width of the bricks being used plus the thickness of a joint. If the arch is to be bonded on the face, there must be an even number of voussoirs each side of the key brick. This will ensure that the springing brick will correspond with the key brick (Figure 58).
3. When the common size of the voussoirs has been determined, complete the arch by drawing the joint lines between the extrados and intrados so that they radiate to the striking point of the curve, thus ensuring that the arch courses are normal to the curve.
4. A template may then be cut from a piece of plywood or hardboard to the shape of each voussoir. This may be obtained by extending two lines of the voussoirs beyond the extrados and intrados of the arch and laying the piece of plywood over them so that the lines extend beyond each end of the plywood. Next, transfer the lines to the face of the piece of plywood by means of a straightedge. The template is then cut off at the lines and should be considerably longer than the depth of the arch extending on the narrow end of the template (Figure 58).
5. Check the accuracy of the template against other voussoirs in the full size drawing, and adjust if necessary.
6. After the shape has been checked, the joint thickness can be determined by laying a straightedge along one of the joint lines on the drawing and placing the

An even number of courses each side of the keybrick

Voussoir lines extended

Template

Semi-circular arch

Figure 58

Template

Stop

Brick to be cut

Diagram showing the use of a template to mark off a voussoir

Figure 59

Temporary piers or 'dead men'

Arch centre

Folding wedges

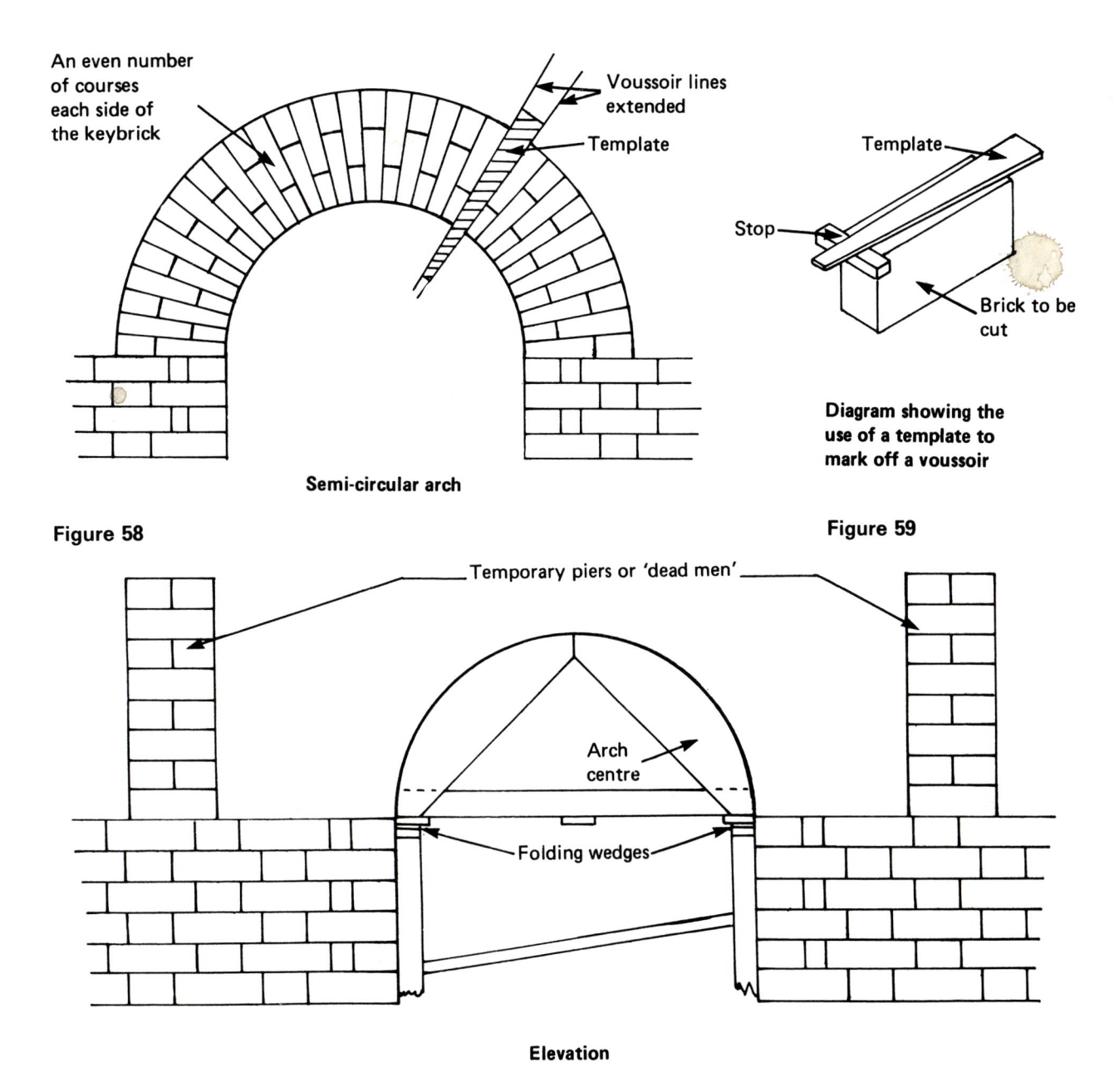

Elevation

Figure 60

template against it so that it completely covers a voussoir. Then slide the template back along the straightedge until the required thickness of bed joint is visible on the drawing. Mark on each side of the template at the point where it coincides with the intrados of the arch. This is the cutting mark and it is helpful if a short length of timber is secured at this cutting mark (Figure 59).

7 Mark the bricks by placing the template on the face of each brick. Mark the soffit of each brick with a square. The voussoirs are then ready for cutting.

8 When the bricks have been axed it is advisable to cut a joggle in the beds to allow for final grouting when the arch is in place.

9 After completing the cutting of the voussoirs, an arch centre is prepared to the shape of the arch and secured in position. A pair of folding wedges placed on the under side of the bearers will allow fine adjustment in the height.

10 With the aid of a pair of dividers mark the positions of the voussoirs (the width at the intrados plus one joint for each course) around the arch centre. This must be set out accurately as the joints must all be equal.

11 Drive a nail in at the striking point or points and attach a length of line so that the alignment of the bricks to the striking point may be checked for accuracy.

12 Check the arch for straightness along its face either by building up the brickwork each side of the arch and stretching a line in between to line up the arch or by building temporary one brick piers each side of the opening (these are called 'dead men') and stretching a line from them (Figure 60).

13 Build up the arch evenly on each side, meeting at the middle or key brick, checking each voussoir for its correct position on the centre and alignment by means of the line from the striking point.

14 When the arch is finally in position the joggles may be grouted with a cement:sand grout poured in the joints.

15 After the arch has been allowed to set, first ease the folding wedges allowing the arch to take its load gently, then remove the arch centre and release the struts.

16 The building of the brickwork around the arch must be done with great care and particular attention given to the cutting of the bricks abutting the extrados, ensuring that there is an even thicknessed joint all the way round the arch. Each brick having to be scribed to the arch curve and then cut and fitted.

Gauged or rubbed work

Many special features in older buildings are often constructed in 'gauged or rubbed work'. This work demands a high degree of craftsmanship and much skill is needed in dismantling and renewing such class of work. The bricks that are used for features of this type are of a very soft, sandy nature which allows for easy rubbing and sawing. However, this characteristic makes it extremely difficult to salvage any units when dismantling these features, as the bed of a brick will tend to adhere to the jointing material and shear away from the body of the brick. Also, if any units are salvaged, it is difficult to clean them, with, for example, a carborundum stone, without reducing the size of the bricks because of their softness. Additionally, since joints are so thin, they do not allow any margin for their re-use after cleaning as the joints then would become oversized.

These bricks must also be handled with great care as the arrises can easily be damaged and it is essential that the bricks are picked up by their beds, keeping one's hands free from the edges of the bricks.

When new bricks are obtained for the renewal of this work, they will be much larger than the normal brick size, and their shape very irregular and they must be 'bedded and squared' before any shapes can be obtained from them. The 'bedding' is carried out by placing the brick on its bed on a york rubbing stone and, grasping firmly with two hands, is rubbed with a circular motion until the bed is even. This is checked with a short steel straightedge or the blade of a steel square. The brick is then turned on to its stretcher face and again rubbed with a circular motion until the face is even but this time, in addition to testing the surface for evenness, the face must also be square with the bedded surface. When these two surfaces are even and square with each other, they are ready for sawing and cutting into any required shape.

Building a camber arch in gauged work

1 Set out the arch full size on a setting-out board for ease of cutting the bricks. This is best done by drawing the skewbacks at 60 degrees to the springing line and extending downwards until they meet, and from this point with the aid of a pair of trammel heads or measuring tape draw a setting out arc with a radius equal to the distance from the meeting point to the crown of the arch. The voussoirs are then marked off on the setting out arc, starting with the key brick and, if the arch is to be bonded, an even number of arc courses each side of the key. The bed joints are then drawn by radiating them down to the striking point. If the arch is bonded then the face joints are drawn parallel to the extrados of the arch (Figure 57).

2 A template is then obtained from the drawing by extending the two lines at the key brick and producing a piece of plywood to the same shape. This is then traversed throughout the whole length of the arch with the aid of two short straightedges. The first straightedge is placed against the skewback and the template placed against it and when it covers a voussoir a setting-out mark is made on the template where it coincides with the setting out arc. A second straightedge is then placed against the template. This template and the first straightedge are removed, the latter being placed against the second straightedge which is also subsequently removed and the template put in its place. This procedure is repeated right through the length of the arch. Any discrepancy in the traversing must be adjusted by planing the template and re-checking by traversing through the arch. This operation must be carried out with care ensuring that the template is accurate because the joints are only 1.5 mm thick and this does not allow any margin for error.

3 The bricks are bedded and squared as previously described.

4 A cutting box having sides equal to the thickness of the arch is then prepared and all the bevels from the intrados of the arch are transferred from the full sized drawing on the setting-out board to the bed of the cutting

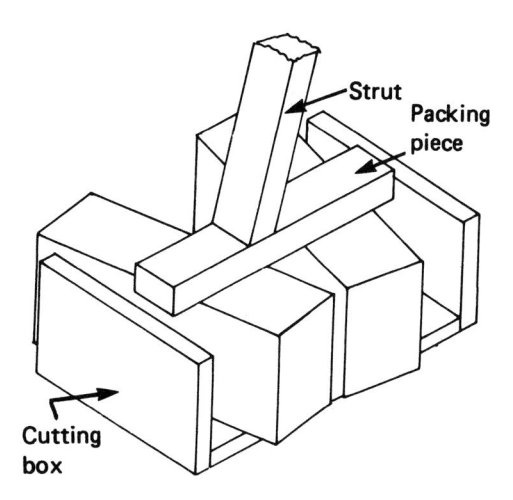

Cutting the angles of the soffits of the voussoirs

Figure 61

Isometric view of a cutting bench made with patent slotted angle iron

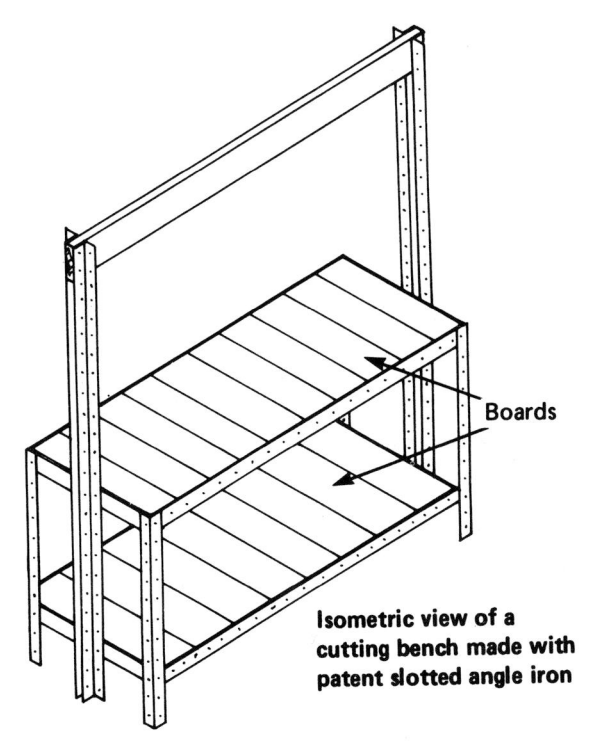

Figure 62

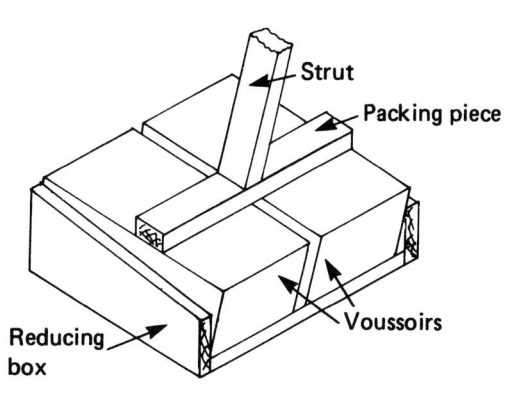

Reducing the bricks to voussoir shape

Figure 63

box. By reversing this bevel both sides of the arch can be set at the same time. This also allows for two bricks to be placed in the box at the same time and sawn off together. The bevels in the box should be numbered 1L 1R, 2L 2R, 3L 3R, and so on until the key brick is reached. The first two bricks, the springers, are then placed in the box on their face sides and placed at marks 1L and 1R respectively, with their ends protruding beyond the end of the cutting box (Figure 61). They are then wedged down from an overhead beam or other similar restraint. If there are a number of such arches to be cut, it is a great advantage to make a cutting bench from slotted angles as shown in Figure 62. The two bricks are then sawn at their tops and ends with the aid of a bow saw. This provides the soffit of the arch voussoirs and reduces them all to the same thickness. This procedure is then repeated for all the voussoirs and the bricks numbered from each side of the arch as previously described.

5   A reducing box is then set up. This may be an adjustable sided box or one which is individually prepared. The sides should correspond to the same shape as the template which was used for setting out the voussoirs on the full size drawing. The soffit bevels are clearly marked on the sides of the box, and these may be taken from the template. The setting-out mark should be clearly marked on the template and transferred to the cutting box. Then each pair or bricks (i.e. 1L and 1R) are placed in their correct positions in the cutting on their beds and reduced to their voussoir widths by wedging down and sawing off with a bow saw (Figure 63). The bricks are now sawn on all faces except the top, and they now have to be cut to their correct length and bevel. This may be done with the aid of a bevel. But a simpler method, and one which is recommended, where there are a number of similar arches to be cut, is to make a jig. This may be formed from a timber base having three shallow sides forming the soffit and the two skewbacks of the arch. The jig must, of course, be exactly the same size as the arch. As each pair of bricks is reduced to voussoir size they are placed in the jig until the key brick is reached. This jig will also check the accuracy of the setting out and the cutting of the voussoirs. When this is done it is a simple matter to mark the tops of the bricks by laying a straightedge at their correct lengths and then placing the bricks in a cutting box and sawing them off to their correct lengths. The voussoirs are now at their correct size and shape.

If the arch is to be bonded then the even numbered courses (i.e. 2L and 2R, 4L and 4R, and so on) may be cut in half at their correct bevel, by using a bowsaw and a cutting box. Alternatively, a groove can be made with the bowsaw at the line of the joint about 12 to 15 mm deep and filled with the same material to be used for bedding

A voussoir joggled ready for setting

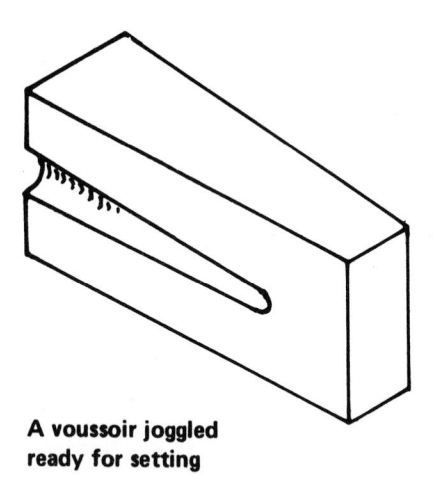

Figure 64

the voussoirs.

6   Joggles may be formed in both beds of each of the voussoirs with the aid of a half round file as shown in Figure 64. As they are finished they should be replaced in the jig which may then be used to carry the voussoirs to the scaffold for fixing. This helps to keep the handling of the bricks to a minimum.

7   When the preparation of the voussoirs has been completed, a soffit board is securely fixed at the head of the opening at the correct height. To facilitate this, folding wedges are placed at each end underneath the board. The wedges will also allow easy striking of the soffit board when the arch has been set in position without causing any undue vibration to the arch.

8   The positions of the voussoirs are then transferred with the aid of a pair of dividers from the intrados of the full size drawing of the arch to the soffit board and clearly marked.

9   A nail is fixed at the striking point of the setting-out curve and a length of line attached for checking the alignment of the voussoirs.

10   The skewbacks are then built up using the line from the striking point to check that they are at the correct slope.

11   The voussoirs are then softly brushed to remove any loose dust and then thoroughly wetted and allowed to drain so that they are just damp when ready for setting in position.

12   The voussoirs are bedded in either a hydrated lime putty or a mix of white lead and shellac. The latter is generally preferable as it forms a cleaner joint and sets quite hard. Unfortunately, however, the white lead is not so readily obtainable today. If lime putty is to be used then this is best placed in a shallow box about 400 mm square and the bricks buttered from this with the aid of a

small trowel (a pointing trowel is very suitable for this purpose). If white lead and shellac are used these may be thoroughly mixed together and thinly spread on the bed of each brick and carefully pressed into place. Whichever method of jointing is used the trowel must not be rubbed along the edge of the brick as it will take off the arris and the joint must be kept as thin as possible, usually 1.5 mm. The voussoirs are laid from each side of the arch until the key brick is placed in position, then the arch is checked for straightness along its face.

13 When all is found to be correct the joggles may be grouted in with a neat cement or a 1:1 cement:sand grout.

14 The arch is then allowed to set, and, when set, it should be lightly cleaned off with a piece of carborundum stone or sandpaper.

Because of the excessive lengths of the voussoirs at the springing courses of a camber arch it is usual practice to bond the courses as shown in Figure 57. The face joints being one third and two thirds up the face of the arch. Thus if the arch is 300 mm in depth the face joints will be 100 mm and 200 mm respectively.

An alternative method that may be used for setting out and cutting a camber arch is as follows. The arch outline is set out full size on a piece of plywood or tempered hardboard. The number of arch courses to be used is determined by allowing an even number of courses each side of the keybrick, then dividing the extrados and the intrados into the same number of equally spaced divisions. The voussoir shapes are obtained by drawing lines connecting these points together. The face joints are then plotted and drawn across the arch at the appropriate courses. Each voussoir is numbered from left and right so that every template will be easily identified. The setting-out board is then carefully sawn along every joint line so that a separate template is obtained for each voussoir.

The bricks are bedded and squared and sawn to their correct thickness as previously described, then the bevels of the soffits and face joints. Also the voussoirs are reduced to their correct shapes by means of the individual templates by placing them in an adjustable cutting box and cutting each pair of bricks, i.e. left and right. The side of the cutting box will have to be adjusted for each arch course. Generally it is better if an arch can be set out on a setting-out curve or curves in order to allow the bricks to be cut without having to resort to the use of individual templates for each voussoir. For example, a pseudo elliptical shape may be obtained by using three continuous curves, as shown in Figure 57. Thus the voussoirs in the haunches of the arch may be obtained from one template and the bricks in the crown of the arch from the second template.

The voussoirs for any gothic arch may be obtained by using one template only. Various types of arches are shown in Figure 57.

Niches

These are often encountered when carrying out restoration work on older buildings and much care and skill are required to replace them. This work is usually carried out in two stages. First the body of the niche then the hood. The bricks must be very soft and easily cut or sawn to shape. Red rubbers are eminently suitable for this work.

The method that may be used for cutting and setting the body is as follows:

1 The plans of the body are set out full size on a sheet of plywood or hardboard and the bonding for each course clearly shown.

2 A template of the inner curve of the body is cut from a piece of plywood or hardboard, and this is used to check the positions of the bricks when the body is being built.

3 Two templates of the shape of the bricks are sawn from a piece of plywood and fixed to the side of a cutting box (Figure 65).

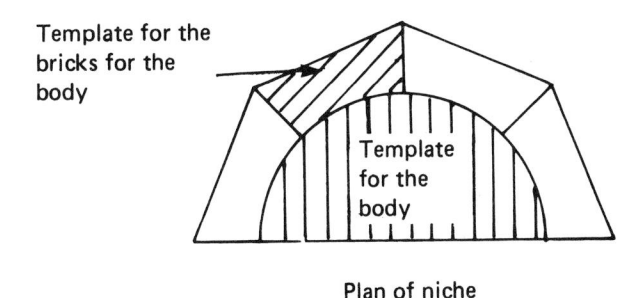

Plan of niche

Cutting box for bricks for the body of the niche

Figure 65

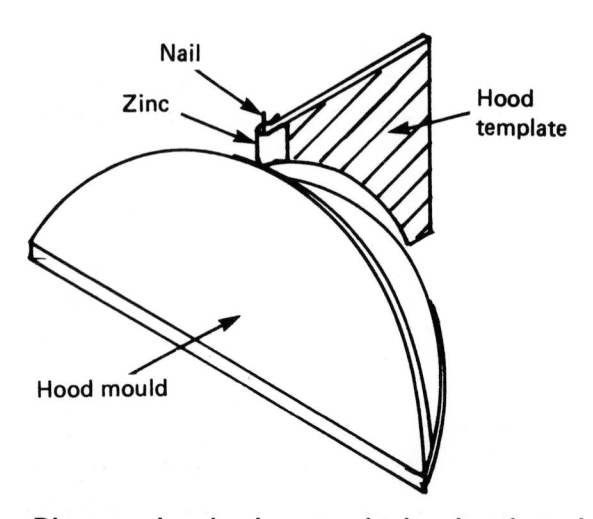

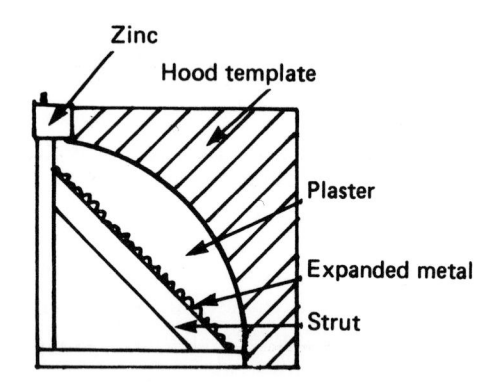

Diagram showing how to obtain a hood mould

Figure 66

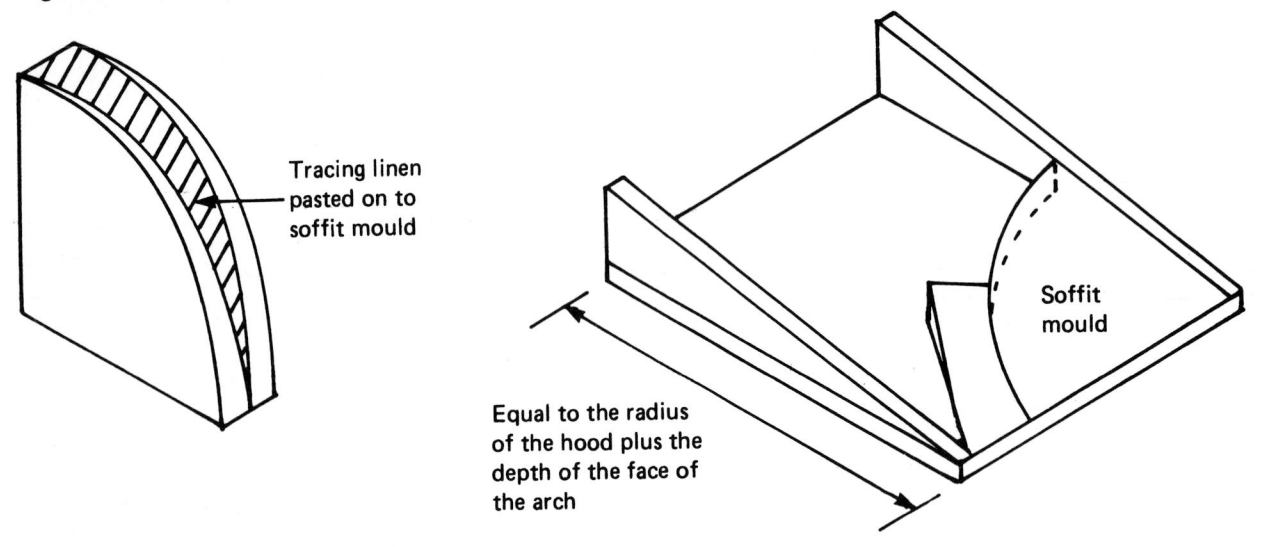

Method of obtaining the soffit mould and its application

Figure 67

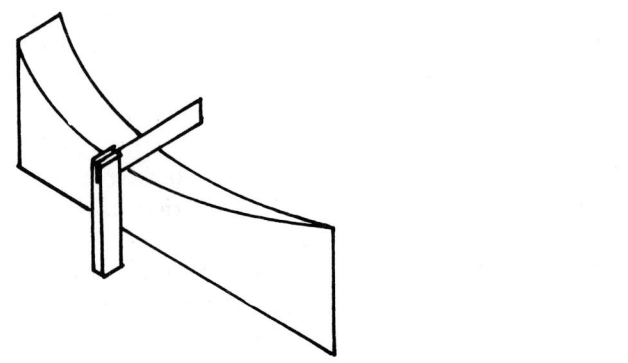

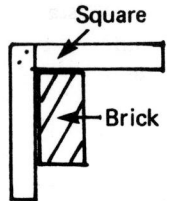

Filing the soffit of the bricks and checking with a square

Figure 68

4 The bricks are bedded and squared and sawn to the required thickness and are placed two at a time in the cutting box, wedged down and sawn to the shape of the templates.

5 When all the body bricks have been sawn to shape they should be well wetted and allowed to drain. They are then built into the body of the niche by using either a lime putty or white lead and shellac. The latter is preferable as it gives a neater and more durable joint. The shape of the niche body being checked at every stage with the aid of the body template, and the bricks kept level at each course, any slight irregularities being removed carefully with the aid of a piece of carborundum before the subsequent course is started. This process is continued until the springing of the hood is reached.

The method for cutting and setting the hood is as follows:

1 The hood must be built on a mould which can be obtained by cutting two pieces of plywood into the same shape as the inside of the niche body. These should then be fixed together at right angles to each other and strutted together to stabilise them (Figure 66). Some expanded metal is then fixed over the struts.

2 A plywood template is cut out of a piece of plywood in the same shape as the convex curve of the niche hood and a small piece of zinc or soft metal cut and fixed at the upper end of the template. A nail is driven into the vertical member of the hood mould and the zinc fitted around it so that the template rotates freely.

3 Plaster is spread over the expanded metal and built up into the shape of the hood mould by rotating the plywood template backwards and forwards over the plaster until it is quite smooth and a true quarter sphere. The surface may be carefully smoothed off with a trowel.

4 When the plaster is set and the mould a true shape, the courses are transferred from the full size drawing to the hood mould. These distances will be equal to the divisions at the intrados of the arch on the face of the niche. When these are carefully checked for accuracy they are then projected right over the mould with the aid of a flexible straightedge or straight piece of soft metal.

5 It is very difficult to cut these bricks to very fine limits and as it is impossible to obtain a pointed unit as drawn on the plans it is, therefore, usual to provide a boss at the centre of the hood. This should also be set out on the hood mould.

6 A soffit mould must now be prepared by cutting a piece of timber which has its width equal to the arch voussoirs at the intrados, and is cut into a quadrant having the same radius as the niche hood.

7 A piece of tracing linen is then placed over any two lines on the hood mould and the two lines transferred to the linen. These should be checked against other pairs of lines on the mould. When it is checked for accuracy, it should be pasted or glued on to the edge of the soffit mould, as shown in Figure 67, and the mould sawn off to this shape.

8 A cutting box is then prepared by fixing two lengths of timber to a plywood base. The shape and length of these timbers are equal to the radius of the hood plus the depth of the arch on the face, and the width equal to one course at the extrados down to a point.

9 The soffit mould is then fixed into the cutting box and the positions of the hood course are marked on to the mould.

10 The bricks are now prepared in the same manner as for the niche body. These are placed in their correct positions in the cutting box and reduced to the required shape with a bowsaw.

11 The upper and lower edges of the voussoirs are checked with a bevel and corrected carefully with the aid of a file (Figure 68). This is the final cutting operation.

12 The hood mould is then placed firmly in position in the niche body.

13 The boss is first prepared by cutting a brick in a semicircular shape with a radius equal to the radius of the required boss. Then the semicircular brick is hollowed out by means of a piece of carborundum stone, file or any other instrument which may be suitable for this purpose. The hollowed portion of the boss should fit snugly over the hood mould.

14 The boss is bedded into place and the hood voussoirs are then carefully bedded around the hood mould working from both sides and ensuring that each course coincides with the course marks on the hood mould.

15 When the hood courses are complete and are set, the next step is to remove the mould. This must be done with great care, as the mould is heavy and can easily slip and chip the bricks which have been bedded in the niche.

16 The final operation is to clean the surface of the niche by carefully rubbing it with a piece of very fine carborundum stone or abrasive paper.

Note: It is most important that great care is taken at each stage in this work to ensure accuracy and neatness.

(a)

(b) Tumbling-in with plinth bricks

(e)

(c)

Overhang to avoid a sharp edge

(d) Tumbling-in courses

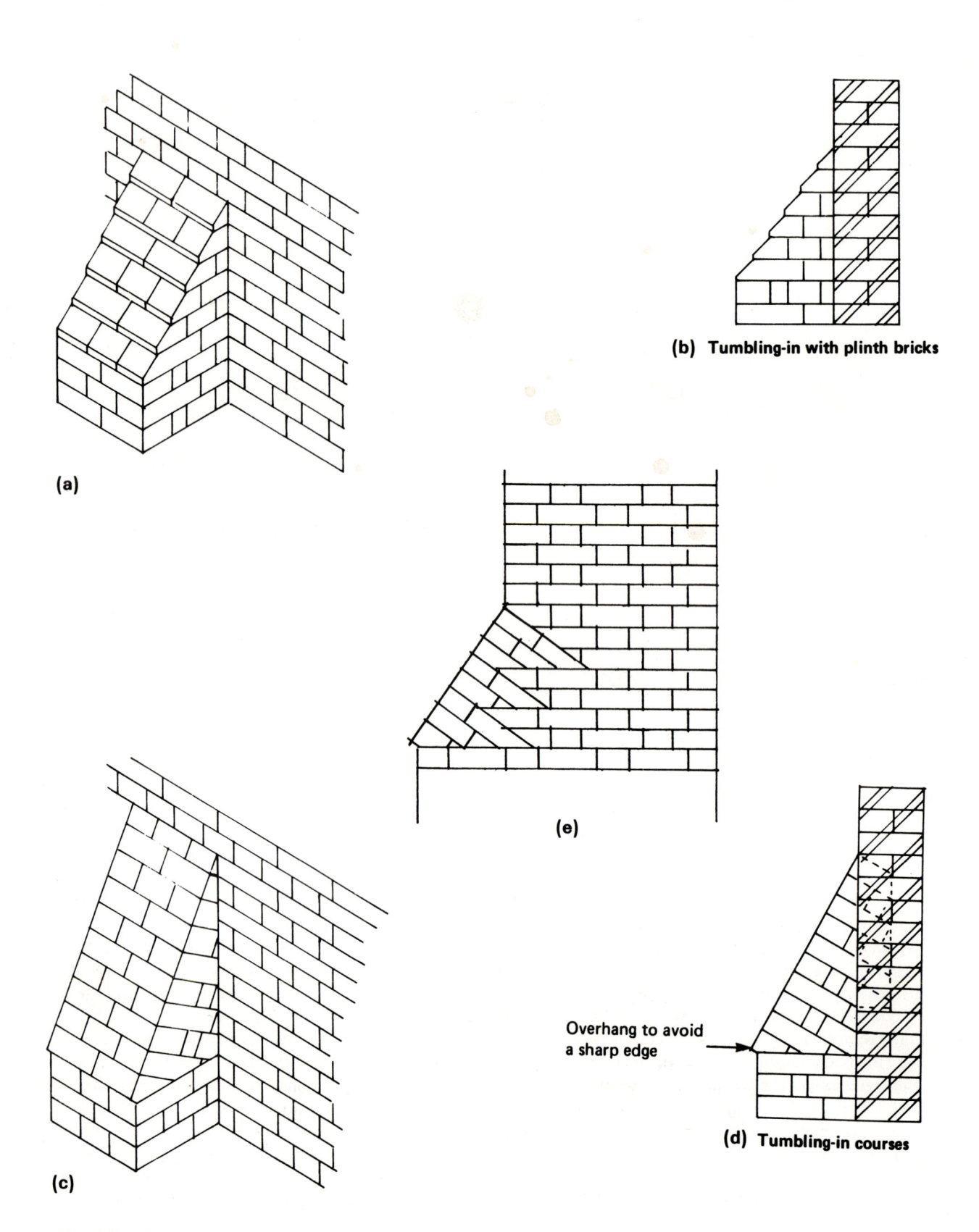

Tumbling-in courses

Figure 69

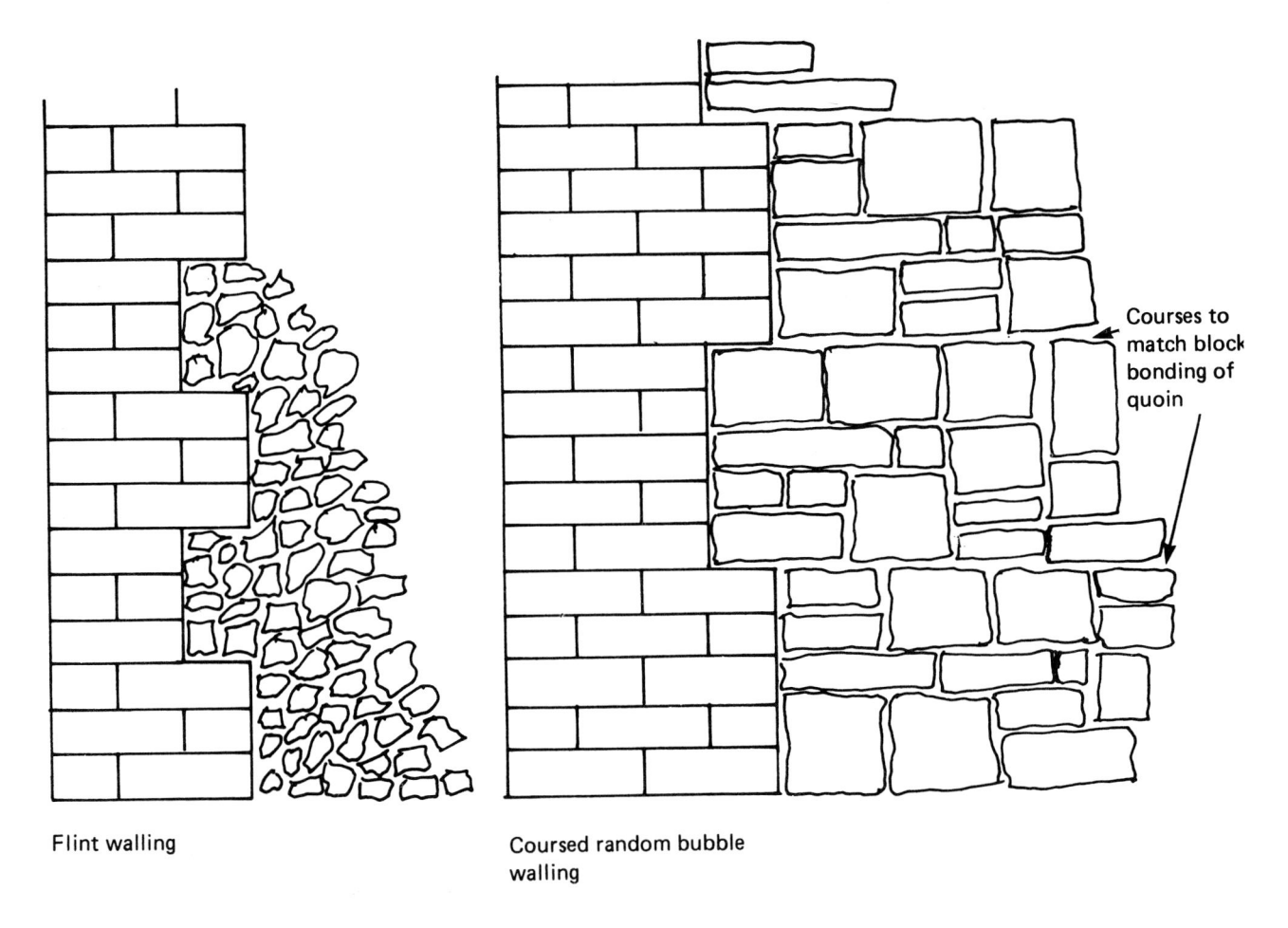

Flint walling

Coursed random bubble walling

Courses to match block bonding of quoin

Brick dressings to quoins

Figure 70

Brick dressings to quoins and reveals

When walls are built with flints, random or coursed rubble it is quite usual to dress the quoins and reveals with brickwork, as shown in Figure 70. Where settlement has taken place the joint between the dressings and the main walling often becomes widened. This is often due to poor bonding between them. Repair of these features usually entails removing the dressings, cleaning the bricks thoroughly for re-use and rebuilding. In order to achieve a good bond between the dressings and the walling reinforcement such as expanded metal should be built in at intervals of 450–600 mm. The reinforcement being taken as far into the walling as is conveniently possible.

Tumbling-in courses

These are used where buttressing walls or piers, or chimney stacks have to be reduced in size. They allow the face sides of the bricks to form the sloping surface to the work, thereby providing a good face to resist the action of rain or frost. The tumbling may be done by using plinth bricks or by laying bricks at an angle with their beds being cut to suit the angle of the slope, as shown in the chimney stack in Figure 69.

When building this work it is usual to overhang the first course of the tumbling-in to avoid having to cut the bricks to a sharp angle since this would leave only a small area to resist the action of frost and rain.

When building the tumbling courses the distance from the springing to the apex should be carefully gauged to avoid introducing a split course which would look most unsightly in work of this nature. It is also wise to plan the cut sections. If the tumbling is only small, the courses may be taken down to the same horizontal course (Figure 69). On the other hand, if the work is quite extensive then the tumbling should be divided into sections given a much more pleasing appearance. These sections should be kept, as far as possible, in similar sizes and shapes so that a reasonable balance is maintained between the horizontal and tumbling courses.

7 Foundations and underpinning

Subsidence

All soils, apart from rocks, are slightly compressible, some more so than others. It therefore follows that all buildings are likely to be subjected to slight settlement when they are completed. But provided that the foundations and the walling are well constructed, little or no damage will occur to the structure since the settlement would normally be expected to be evenly distributed all round the building.

However, differential settlements may occur during the life of a building, and this may lead to various degrees of cracking. These settlements will be caused by changes in the bearing pressure of the soil which can be affected by:

1. Frost heave in chalk and silty soils.
2. Variations in ground water levels which may be due to excavations nearby causing shrinkage in the foundations.
3. The sub-soil becoming wet due to excessive rain or flooding, causing expansion.
4. Unforeseen hazards in soils, such as unsuspected or inadequately consolidated made-up ground.
5. Local vibration.
6. Swallow or sink holes in chalk or limestone areas.
7. Slips on sloping sites.

If severe movement has taken place it may be considered necessary to take remedial measures. First a careful survey should be undertaken to trace the source of the problem. This may be carried out by digging a trial hole at the base of the wall (if it is required to be taken down to any great depth then timbering must be used to stabilise the sides of the excavation). The following factors can then be determined:

1 To identify the type of foundations that were originally used under the structure such as:

(a) Traditional strip foundations consisting of a concrete slab with brick footing courses.
(b) A concrete strip without brick footings.
(c) Concrete trench fill.
(d) If the building is very old then it may have no concrete strip at all or has the walls built of elm baulks.

2 The depth of the foundation below the adjacent ground level.
3 The table level of any ground water.
4 The type of soil in which the foundations are built. If this is a clay soil then this should be inspected in close detail and, if necessary, soil tests should be carried out to ascertain its composition and behaviour regarding drying shrinkage, cohesion of its particles and bearing pressure under wet, damp and dry conditions. Careful note should also be taken to discover if there are any stiff fissures or tension cracks in the clay which will be a clear indication that considerable movement has taken place, or is taking place, in the soil. If such cracks are in evidence then it would be advisable to carry out investigations to at least a depth of 3 metres or more in order to attempt to ascertain the extent to which the cracks have penetrated. The size of the fissures will indicate the amount of movement that has occurred in the soil at varying depths.

The fissuring effect may also cause great variation in the shear strength of stiff fissured clays which, in turn, can cause defects in the structure. Usually the fissures get narrower as they penetrate deeper into the soil. Also, the movement, due to shrinkage, diminishes as the fluctuation in the level of the surface water decreases. Some idea of the depth of a fissure may be gained by carefully inserting a wire probe until it reaches resistance, but this should only be taken as a guide and not a detailed inspection.

5 To determine if the concrete foundation has sheared or tilted due to uneven shrinkage. This may be caused, for example, by evaporation of the surface water taking place on the external side of the structure whereas none occurring on the inside of the building, or if it has moved horizontally sideways due to differential horizontal forces such as those which may occur on sloping sites or where the sub-stratum is sloping. There may also be sideways movement in concrete trench fill foundations due to the differential moisture content on opposite sides of the trench fill which causes a complete loss of adhesion on the external face of the concrete while maintaining adhesion on the internal face. This produces a damaging eccentricity of reactive horizontal forces which may cause the wall to slide outwards.

6 The position of trees and large shrubs and their types. In clay soils the close proximity of trees will accentuate any movement in the soil due to shrinkage and swelling. It is estimated, as a rough guide, that they should not be situated nearer to the building than a distance equal to their height. The rate of moisture absorption will vary with different types of trees – quick growing trees, such as the pines or conifers, require a great deal of water for their growth and foundations are, therefore, specially vulnerable to damage when these trees are situated close to the structure.

In the cases where defects in the foundations can be attributed to the presence of trees, the trees can be removed but this may cause the soil to start swelling due to the moisture not being extracted from the soil. If it is possible in such cases, the soil should be allowed a period to maintain reasonable moisture equilibrium before carrying out major remedial structural work in the foundations.

7   The position, depth and width of cracks should be determined and recorded. Any current movement should also be checked by using 'tell tales' fixed in such a manner that they bridge the cracks. Thin pieces of glass may be bedded across the cracks and these will give a rough guide to the amount of movement. A more precise method of investigation, however, may be made by plugging two short lengths of angle iron to the wall, one each side of the crack and securing a dial gauge between the two angle strips. Frequent readings will give a clear indication of any movement and also measure the amount of it. It is wise to fix a cover plate above the dial gauge to protect it from the weather.

Interesting readings may be noticed on cracks which have occurred in buildings which are situated adjacent to tidal rivers or the sea shore. It is quite usual for regular movement to take place coinciding with the rise and fall of the tide. In all of these surveys it can sometimes be helpful if the history of the site, or any movement of other buildings in the vicinity, can be obtained from local records.

Remedial work

If the results of a survey indicate that movement under the foundations is likely to continue and cause further damage to the structure, measures must be taken to prevent such settlement or movement by extending the foundations to a relatively stable soil and below the zone of seasonal changes. Such remedial work may be carried out by:

1   Underpinning the existing foundations.
2   Using one of the specially designed piling systems.

Underpinning

This is generally the most suitable method of stabilising the foundations but it is extremely important to remember that much care and thought are essential if this operation is to be successfully carried out. The procedure of working is as follows:

1   The cracks in the wall are carefully examined and recorded, and 'tell tales' fixed at strategic points over the cracks to monitor any further movement should it occur during the underpinning operations.

2   The wall is then divided into a convenient number of sections, generally not exceeding 1.5 metres each depending upon the condition and stability of the walling. These working bays should have sufficient room to enable a man to work inside them.

3   A batten should be fixed along the face of the wall throughout its length. This batten will act as a datum and, therefore, it must be fixed perfectly level. All levels of the extended work will be measured from this datum line.

4   A sequence of working must then be arranged. This will often be determined by the conditions on the site and the accessibility of the working area adjacent to the walling. No two adjoining bays must be excavated at the same time. If it is impossible to avoid consecutive bays being worked in then it is essential that the first bay is fully completed and allowed to set hard before the next section is started. Also it is important to remember that any unsupported length of the wall should not exceed at any one instance one quarter of the total length of the wall during underpinning operations.

Typical sequences of operations which allow continuity of working are shown in Figure 71.

5   Each working bay is excavated in turn, usually by hand. The earth in front of the wall being removed first to allow the wall to retain support for as long as possible. When a firm base is reached, the soil underneath the foundations is excavated, taking care not to extend too far at the back. The deeper that the excavations are taken, the more filling must be done, so it will be more economic to keep the earth as tight to the new walling as is conveniently possible.

6   When the level of the new foundation is reached the width of the new foundation is marked out and the ground excavated and the bottom thoroughly levelled and rammed.

7   The projecting arm of the existing concrete may be cut off if required but often this is left intact so the whole of the foundation is completely underpinned.

8   The new foundation concrete is laid in the trench to the required thickness and at the correct level measured down from the datum line.

The concrete must have adequate strength ultimately to resist the loading that it is to carry and also to be resistant to the action of any sulphates which may be present in the soil. A suitable mix for the concrete would be 1:2:4, giving sufficient strength for general work of this nature. If a higher strength concrete is required, this should be designed specially for the purpose.

The aggregates should be correctly apportioned, preferably by weight rather than by volume, the cement added and all thoroughly mixed together. The correct volume of water may then be added to the mix and again thoroughly mixed together. The amount of water added to the mix is very critical since too much will seriously affect the strength of the concrete and too little will not allow the mix to be soundly compacted. This will affect the strength and durability of the concrete. Generally, if the compaction is to be carried out by hand then the total amount of water (including that which is contained

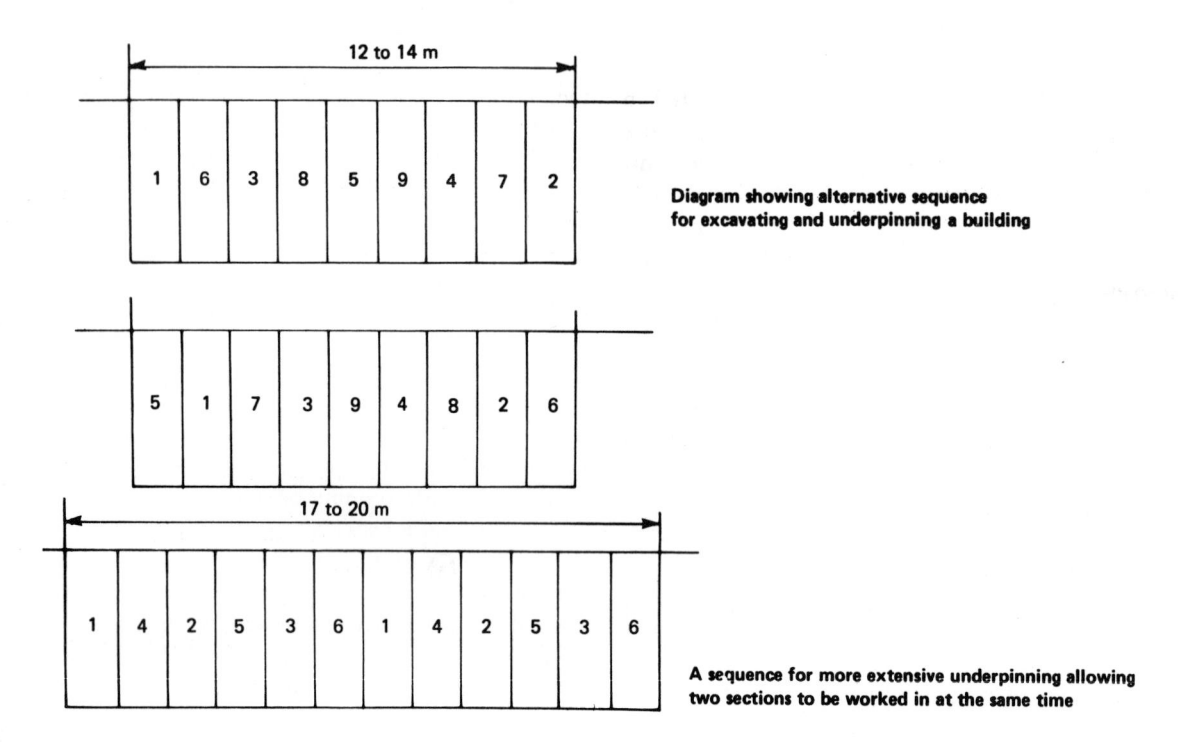

Figure 71

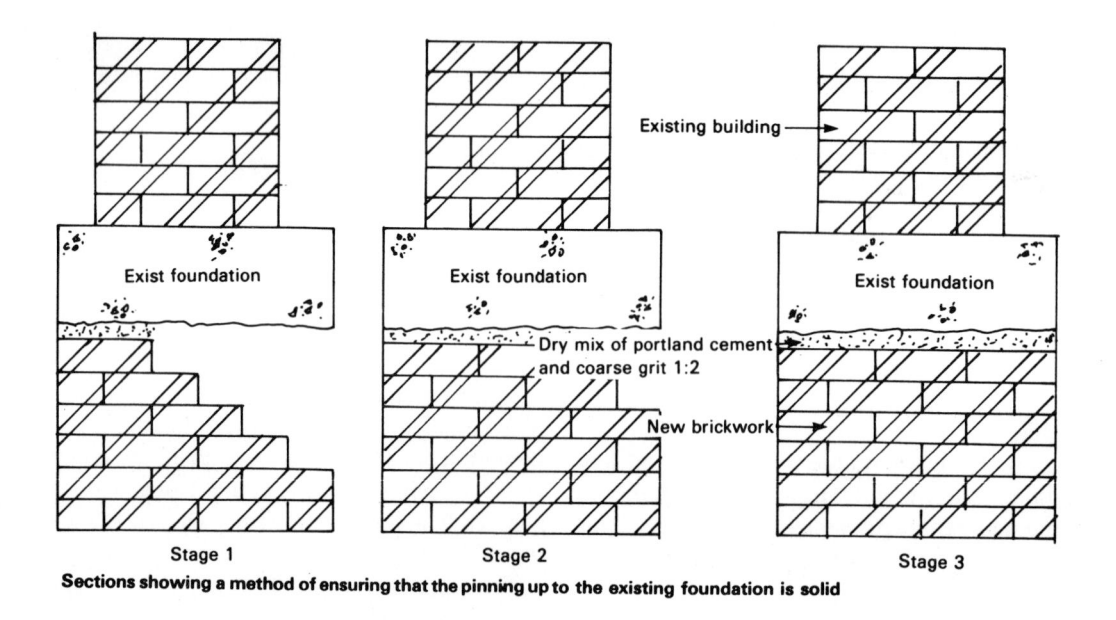

Diagram showing alternative sequence for excavating and underpinning a building

A sequence for more extensive underpinning allowing two sections to be worked in at the same time

Figure 72

Sections showing a method of ensuring that the pinning up to the existing foundation is solid

within the aggregates) should be about 0.6 of the weight of cement, i.e. 30 kg or 30 litres of water to every 50 kg of cement. If, however, the compaction is to be done with the aid of machinery then this water content should be reduced to 0.5 of the weight of cement.

The concrete must be thoroughly compacted and then allowed to set and harden. Rapid hardening cement can be used to great advantage in this work as it allows the work to proceed without delay and avoids leaving the ground underneath the building from remaining open to the elements for too long a period and causing soil shrinkage through drying out, or expansion if subjected to rain or water penetration. If the soil is suspected to contain sulphates in any substantial quantities, the concrete should be mixed with sulphate resisting cement. If reinforcement is required to be laid in the foundations it should be bent up at the ends, so that when the concrete in the adjoining bay is laid it will be bent down and incorporated into the concrete. Any water in the excavation must be pumped out before laying the concrete.

9 When the concrete has set and hardened the brickwork is built up to the underside of the old foundation. If the brick pier is an end, it will be toothed at one end facing the centre of the walling. If it is a section in the centre of the wall it will be toothed at both ends. This will allow the brickwork to be well bonded together forming a solid mass below the old foundations. The space behind the wall and earth must be filled in and solidly compacted.

The brickwork is then taken up to within about 50 mm of the old foundations and the space between thoroughly filled and compacted with a dry stiff mix of fine concrete. If the walling is very thick then this filling may be done in stages as shown in Figure 72. Special care however must be taken with this packing as the stability of the old walling will largely depend upon the degree of compaction that is achieved.

10 This operation is then continued in the other working bays following the pre-arranged sequence and the new concrete foundations carefully linked with each other and the brick piers bonded into one another.

Note: It is wise to remember that the working bays are quite separate from each other. Therefore to ensure that the work is soundly bonded together each bay must be carefully gauged down from the datum line, and a gauge rod marked in the appropriate courses can be used to great advantage with this work.

As with all underpinning work it is most important to work accurately and carefully, without undue haste, and to allow ample time for the work to set and harden before allowing it to take any loading from the structure.

Piling

There are a number of specialist companies who design systems for underpinning, each having its own particular advantages and disadvantages. One method that has been developed by the Frankipile Co. Ltd, is a very efficient system for carrying out work of this nature and is recommended for use where the work is required to be done without causing any vibration to the structure.

The pile is formed in precast concrete sections successively jacked hydraulically into the ground until a pressure gauge indicates the predetermined bearing capacity. Each section has a steel-lined hole running through it. The steel lining assists in the locating of the sections and ensures that they are all in line. The first has a pointed steel toe-piece to make penetration in the ground easier.

The process of installing these piles is as follows:

1. A hole is dug below the existing foundations of the structure.
2. The first precast concrete section, which is the pointed one, is placed into position below the foundations.
3. A bearing plate is then positioned between the jack and the foundation.
4. The jack is then put between the first section and the bearing plate. The first section is forced downward by the action of the hydraulic jack which is powered by a pump situated outside of the excavation.
5. When the top of the first section is almost flush with the ground the jack is removed and the process repeated with the second and subsequent sections.
6. As each section is added a length of steel tube is inserted into the hole and grouted into position to make an effective joint between the sections.
7. The operation is continued until the pressure gauge indicates sufficient penetration resistance to ensure adequate bearing capacity.

Structural repairs

When the foundations have been adequately stabilised then the repair work to the walling of the superstructure may be started. Severe cracking should be cut out and rebuilt as described in previous chapters and thin cracks repointed. If damp-proof courses have been damaged then these may be replaced by cutting out the existing damp-proof course and the course of bricks immediately above it in sections of about a metre or so, replacing the damp-proof course. (A lead lined mineral felt would be very suitable for this work as reasonable lengths can be kept in a roll and unrolled as the work progresses.)

Each section is built up solidly before proceeding to the adjacent section of the walling. The procedure is repeated until the damp-proof course is completely replaced. To avoid cutting out adjacent sections this work may be carried out from each end of the wall but the total length of unsupported wall must not exceed a quarter of the length of the wall. Also, a sufficient length of time must be allowed for each completed section to set and harden before commencing the cutting out for the adjacent section.

Glossary of terms

Aggregates These are divided into two main groups (1) *fine aggregate* or sand which passes through a 5mm sieve and (2) *coarse aggregate* that which is retained on a 5mm sieve and is mainly used as a filler in concrete.

Arches are a way of bridging openings and are divided into three main groups (1) *rough axed* in which the joints are often wedge shaped and not the bricks and are used on work which does not require a high standard of finish. (2) *fine axed* arches which are carefully set out with the voussoirs all cut to the same shape and laid *normal* to the arch curve and (3) *gauged* arches which are very ornamental and expensive as the bricks are of a special type being soft and sandy allowing them to be rubbed to shape rather than cut, and require much detailed preparatory work. Such bricks are bedded with a very fine joint, such as neat white lime or a mixture of white lead and shellac; the latter is preferred as it gives a more even and durable joint.

Attached piers For economic reasons some walls are built with thin walling and it is, therefore, necessary to strengthen them by thickening the wall at intervals with buttresses or attached piers. For these piers to be completely effective they should be taken up to within a distance from the top of the wall equal to three times the least thickness of the wall. The tops of the piers should be protected against the effects of the weather. This is often done with *tumbling-in* which also provides a decorative finish. The piers should be constructed off the same foundation as the wall.

Approved Inspector When submitting an application together with accompanying drawings and specification for a proposed new structure, the Technical services Department will carefully check the details to ensure that they comply with minimum standards recommended by the Building Regulations. Such checking may also be carried out by an *Approved Inspector* who may be an Architect, Chartered Builder, Control Officer, Engineer or Surveyor who is in private practice.

Base in relation to a wall means the underside of that part of the wall which rests immediately upon the footings or foundation concrete.

Basket weave pattern usually comprised of sets of bricks laid alternately vertically and horizontally to provide a decorative panel.

Battering Face This is the term given to walling which is built leaning back; this is particularly useful for building retaining walls or boundary walls in order to achieve greater stability.

Bench mark This is a fixed point on the earth's surface and has a known level above the Ordnance datum of *mean sea level*. The *datum level* on a site is often taken from the nearest bench mark and a *fixed datum peg* set up and from which all the heights and site levels are then taken.

Block Bonding commonly used at the corner of a building to give a decorative finish; the blocks may be built with contrasting bricks or other materials. Block bonding may also be used instead of toothing every other course to extend or thicken a wall or tying a partition wall into a main wall. Blocks of three or more courses may be used to provide stability of the walling.

Bonder Course This is a course of sufficient depth and thickness to transfer the weight of cladding walling to the structural frame of the building.

Bonding of Brickwork has the prime function of distributing the structural loading evenly over the whole of the walling. It also provides decorative features by the introduction of various patterns to relieve the monotony of the walling.

Broken Bond When setting out facework in good quality brickwork due note should be taken of the position of door and window openings and the bond should be set out on the first course. This often means that true bond cannot be maintained between the reveal perpends so a broken bond such as a threequarter brick or other pattern has to be introduced.

Buttressing pier may project from one or both sides of a supported wall.

Buttressing wall means a wall which is constructed so as to afford lateral support to another wall from the base to the top of the wall.

Capping is used to give protection to the upper surface of walling against harmful effects of the weather. It also provides a decorative finish to the walling.

Centre During the construction of an arch the voussoirs are temporarily supported by a *centre* which is made up from a number of small section timber members assembled into the shape of the arch.

Chimney is defined in the Building Regulations as any part of the structure of a building forming any part of a flue or passage for conveying the discharge of a heating appliance to the external air.

Composite walling are walls which are built with differing types of bricks or different thicknessed bricks on opposite faces. This is also applicable to stone faced buildings which are backed with brickwork or walls built of brick and blocks.

Concentrated loading occurs when a load such as a girder is applied at a single point on the walling thus the wall has to resist crushing and bending. In such cases it is usual to rest the beam on a template or padstone to distribute the load evenly.

Copings are used to give a decorative finish to parapet walls; also to provide a protection on the upper surface against the effects of the weather. Such copings include *brick-on-edge*, *brick-on-edge and tiles*, *saddle back* and *feather edge* types of construction.

Corbels sometimes referred to as *oversailing courses* are a series of bricks which may be used to thicken or lengthen a wall to form a decorative feature. In no case should a corbel overhang a distance greater than the thickness of the wall measured immediately below the corbel.

Cross walls are provided to divide a structure into compartments also to provide stability within the structure.

Crown is the highest point of an arch at which the key brick is placed.

Damp-proof Course provides a barrier to the passage of moisture from the ground into the structure. Where buildings are constructed below ground level it is then necessary to insert vertical as well as horizontal damp proof courses. They are also introduced in parapet walling to prevent the passage of rainwater downwards.

Datum is the basic level on a site from which all the various heights in the building are taken. The level of the datum is often determined from the nearest Ordnance Survey bench mark or if this is not available from another adjacent fixed level.

Dentil Courses are formed by projecting alternate headers in oversailing courses they are often introduced to provide a decorative feature at the head of a wall. If these headers are laid at 45 degrees to the face of the wall then these are referred to as *dog toothing*.

Drain is the length of pipe to convey soil water or waste water from a building or buildings within one boundary the owner of which is responsible for its maintenance and repair.

Dry area When a building is taken below ground level it is often common practice to provide an open space between the building and the soil, this is also referred as a *basement area*. The main function of this dry area is to allow daylight to be provided to the basement of the building.

Drying Shrinkage is the shrinkage or movement which takes place when materials, which have been mixed with water, dry out. When building units such as bricks or precast blocks become wet they expand and conversely when they dry out they will shrink; this action is referred to as *moisture movement*. It is most important that this movement is well understood and allowed for in a long length of wall by providing *construction joints* at intervals along the wall in order to control this movement and prevent unsightly cracking which may occur in the wall face if such joints were not provided.

Effective height If a wall or column is loaded without eccentricity its loading capacity is dependent upon the individual strength of the bricks, the strength of the mortar and the slenderness ratio (which is basically the ratio between the height and width of the column). This may cause a problem for many complicated wall systems but can be overcome by adjusting the measurement of the height of the wall – this being called the *effective height*.

Efflorescence is caused when bricks contain soluble salts such as *sodium sulphate* (Glauber Salts) or *magnesium sulphate* (Epsom Salts). The action of rainwater will dissolve these salts thus forming a solution. When this solution dries out the water will evaporate into the atmosphere leaving the salts on the face of the wall in the form of crystals. If a concentration of these salts should form behind the face of the brick then there is always a possiblity of *spalling* of the brick faces occurring due to the expansion of the crystals.

Extrados is the upperside of an arch as seen in elevation.

Fair faced brickwork or *neat work* is walling which is built with facing bricks and the joints *jointed* or *pointed*.

Fire brick lining is the lining of a chimney with refractory bricks to form a flue.

Flaunching is the rendering to form the flue over a fireback in a fireplace. It is also the term for rendering around a chimney pot to form a weathering to protect the top of the chimney stack against the effects of the weather.

Footings are the brick courses built at the base of a wall immediately above the foundation concrete.

Foundation is the concrete structure at the base of a wall to distribute the loading of the building over the ground immediately below.

Frost heave can occur in some types of soils especially silts and chalks which are liable to severe expansion when their water content freezes. Foundations built upon these soils should be at least 1.25 metres deep in areas subject to severe frosts.

Gable shoulder or *Springer* is the corbelling at the bottom of a gable end and often made into a decorative feature of brick tiles or stone.

Gauged mortar is a mortar in which every mix is measured in correct proportions so that they are of uniform colour, and mixed on clean areas or on properly constructed platforms either by hand or with the aid of a machine such as a *mortar pan* or *concrete mixer*.

Haunch is the lower part of an arch from the springing line to midway to the crown.

Herringbone pattern consists of a series of patterns of bricks which are laid at 90 degrees to each other but at 45 degrees to the horizontal plane. Such patterns may be laid vertically or horizontally and are mainly used in decorative panels.

Honeycomb walling are dwarf walls built on top of the oversite concrete to carry a timber floor. These walls are built with holes through them to allow free passage of air underneath the floor.

Intrados is the underside of an arch when seen in elevation.

Jamb is the surface of brickwork which forms the reveal of a window or door opening.

Joggle joint After the voussoirs of an arch have been cut it is usual to form a deep groove in both beds. When the arch is set into place and checked for accuracy then the joggles are filled with a strong mortar grout to form a solid key.

Jointing is the term given to finishing the joints of fair faced brickwork as the work proceeds.

Junction wall or *buttressing wall* is a wall built at right angles and bonded into another wall.

Key brick is the highest or central voussoir, usually the last to be built in an arch.

Label course is an extra course built around the extrados of an arch usually built with projecting headers giving protection to the arch against the weather. It also gives greater depth to the arch without having to cut shaped voussoirs.

Lining up a gable A temporary profile is erected in line with the roof and immediately behind the gable wall to be built. A line is secured from the top of the profile to the gable shoulder and the gable wall is then cut and built to the line.

Manholes or *inspection chambers* are generally constructed of one brick walling at intervals in a drainage or sewer system to allow access for cleaning out in the event of a blockage occurring in the system. They also provide access for junctions and any changes in invert levels to be made. It is also necessary for provision to be made for a sound foothold at the bottom to allow a person to stand safely. A manhole should also be sealed with a suitable iron cover at the top.

Mortar is a mixture of a matrix such as cement, lime or masonry cement and sand. The proportions of mix depends upon the strength required in the mortar but in no case should it be stronger than the bricks which are to be laid on it. The mortar should be workable so that it can be handled easily and have a good bond with the bricks or blocks. It should also be durable and resistant to the effects of the weather.

Natural bracketing is the bridging of a temporary opening in a wall which has been left to allow people to pass through while the building is being constructed. The bricks are corbelled over with the bond of the brickwork.

Oversite concrete is the bed of concrete which is laid over the whole of the area within the external walls of a structure. Its main function is to prevent rising damp and to exclude the growth of plant life underneath a floor.

Plinth bricks are special purpose made bricks which are used in courses laid to reduce the thickness of walling above and to provide a weathering to the plinth courses.

Pointing is the process of finishing off joints after the wall is completed. The joints are raked out at the end of each day's work and when the whole of the work is completed the walling is cleaned down, the joints filled with gauged mortar and finished off with a tooled or trowelled finish.

Projecting arm is the distance that the foundation concrete projects from the face of the brickwork.

Quoin is the corner of a building.

Reveal is the face of brickwork forming the vertical edge of a window or door opening.

Reverse bond is where the bond differs at each end of a course of brickwork on a pier e.g. if there is a header at one end then there will be a stretcher at the other.

Sand courses Where pipes are intended to pass through a wall *sand courses* may be introduced as a temporary measure to allow the work to proceed. After completion of the walling the bricks laid in the *sand courses* can then be easily removed, the pipes laid and the hole in the wall built in solidly.

Separating wall is a wall which is common to two adjoining buildings.

Serpentine walling is walling which curves in and out along its length and is normally confined to boundary walling. Also known as 'crinkle crankle' walling.

Skewback is the sloping abutment from which an arch springs.

Soffit is the surface of the underside of an arch.

Soldier arches or *brick lintols* are flat units bridging an opening but because of their low tensile strength rely on another means of support. This may be provided by reinforcing rods passed through perforated bricks or an *in-situ* concrete lintol placed behind the brick lintol with wall ties or similar items being placed in the vertical joints of the lintol and bedded into the concrete lintol.

Spalling This is the term given to the breaking down of the surface of a brick face due to the action of frost upon the water content of the brick, turning it to ice and forcing the surface skin off the brick. Dense hard burnt bricks are not usually prone to this but it is inadvisable to use fletton bricks in damp exposed situations as they are liable to spall. High concentrations of efflorescent crystals forming behind the faces of bricks may also cause *spalling*.

Springing line is a horizontal line drawn through the lowest points of an arch from which the arch curve starts.

Squint bricks are special purpose made bricks which are used to build obtuse angled corners.

Stepped foundations are introduced on sloping sites in order to economise on the excavation work. Each section must overlap the one immediately below by at least 300mm or the thickness of the foundation concrete whichever is the greatest.

Striking point is the geometrical centre point from which an arch curve is drawn. The voussoirs should radiate to this point thus ensuring that they are *normal to the curve*.

Struck joint is formed by rubbing the trowel along the joints of the brickwork as it is being built – the lower edge of the joint being struck back.

Sub foundation or *ground bearing* is the soil immediately below the foundation concrete. This should be well rammed before the concrete is laid to ensure that there is no loose material. Also the concrete should be laid as soon as possible after the excavation is completed in order to prevent the soil drying out and being affected by *drying shrinkage*.

Supported wall is the wall which receives support from a buttressing wall or pier.

Template or **padstone** is a unit of stone or concrete bedded on a wall to receive a girder or truss in order to transmit the load evenly upon the wall surface.

Tile creasing are courses of creasing tiles which are plain flat tiles without nibs as in the case of roofing tiles, and laid as cills at window openings or as decorative features under a brick-on-edge capping to a wall.

Toothing is the indentation at the end of a wall when the structure is required to be extended at a later date. The indentations are formed according to the bonding of the brickwork in the wall.

Trammel is a length of thin batten cut to a suitable length and used to set out curved work.

Tumbling-in The top of a buttressing pier may be finished off by turning the top of the pier into the main wall so that a weathering is provided. Also a decorative feature is effected.

Underpinning is made necessary when it is required to lower the foundation of a structure and is work which must be carried out with great care to ensure that the stability of the structure is safeguarded.

Voussoirs are the individual wedge shaped bricks or stones used to form an arch curve.

Weather struck joint is the finishing off of a bed joint by rubbing the trowel along the mortar and striking back the upper edge of the joint.

Wheel arch or *bulls eye* is a type of arch used around any opening which is in the form of a complete circle.

Working along a line When the corners or profiles are erected a line is stretched tautly between them and the bricks on each course are laid to the line, thus ensuring that the courses are straight along the face of the wall; also that they are laid horizontally.

Working drawings show the details of the work to be carried out. Whilst these are drawn to scale nevertheless preference should be given to figured dimensions when setting out the actual work. The architect's attention should be drawn to any obvious errors should they occur.

Student exercises

The following exercises have been compiled to assist the reader in continuing his researches into the diagnosis of faults which may be found in old brick structures which are in need of repair. Exercises numbered 1 and 2 illustrate how a systematic approach to the problem may be used. Each problem will, however, need its own particular methods.

1. Extensive repair work is to be carried out to an existing building and a number of facing bricks are required for the external walling;
    (a) What Association would be helpful in trying to locate a suitable brick manufacturer?
    (b) Having obtained a supplier what information would he be likely to require from you?
    (c) What would be the minimum length of notice for supply he is likely to require from you?
    (d) If any specially shaped or moulded bricks are required how would you give full details of your requirements?
    (e) If any special shaped bricks for units such as arches are required how would you convey your needs to the supplier?
2. An external wall of a three storied building is found to be bulging and is giving cause for concern for the safety of the structure. The walls are of solid construction and apart from the bulging appear to be in a reasonable condition.
    (a) (i) How would you check this wall to ensure its stability?
        (ii) What factors should be investigated and taken into account when considering what repair work is to be carried out?
        (iii) What methods may be used to prevent further movement of the wall?
    (b) If the wall is found not to be in a reasonable condition and there appears to be considerable damage, what action should then be taken to carry out repairs to the structure?
3. A large extension is to be built on to a building which was erected in 'the thirties'. The facework of the existing building was built with hand made sand faced reds bedded in a mortar which is fairly dense and is a light creamy colour. All of the facework is in good condition. The new extension will be attached to the old building and will be clearly visible when the work is complete. The architect who has been commission to carry out the work of the new extension is adamant that the new work must have a good match with the existing building. He has managed to locate a supply of bricks similar to the original and is now concerned with the mortar. What instructions should he issue to the contractor before commencing the work also during its construction in order to ensure a reasonable matching of the new work with the old?
4. After a detailed investigation has been carried out on an old building it is discovered that the external walling was built with hand made facing bricks but in recent years the walling was rendered with a fairly soft mortar which has not weathered very well and shows signs of considerable damage. It has now been decided to remove the rendering and restore the face brickwork so describe the method of carrying out this work and explain how defective bricks may be replaced economically.
5. A building which is about twenty five years old has a parapet wall which was built in brickwork with a stone capping and rendered on its internal face. This rendering is now found to be badly cracked in rectangular patterns and has become detached from the wall in several places. State the likely cause of these defects and the repairs that should be carried out to remedy them.
6. A chimney stack serving a coal fire in an old building is noticed to have a distinct leaning and you are asked to investigate this structure. A scaffold is erected and when inspection of the stack is permitted what defects would be expected to be seen? How would they be caused? What repair work is likely to be needed to put the stack back into a sound state.
7. A large mansion is to have extensive restoration work carried out to the external walling. Describe how a detailed survey of the walling may be carried out. State how full details of the repair work to be carried out, would be compiled in a systematic manner.
8. An old building is found to have a considerable amount of lichens and moss growing on the face of the wall both internally and externally. What steps may be taken to eradicate this growth and what other action do you think might be necessary to prevent further growth?
9. A new structure is to be built immediately adjoining an existing building. The foundations of the new structure will be several metres below the existing building. Describe a method that may be adopted to ensure the stability of the old building before the new foundations are laid.
10. An old building has gauged camber arches over its window and door openings; several of these have become damaged through settlement and other reasons. Describe how the damaged arches should be removed and

replaced with new units to match the existing arches.

11. The face of a building is to be re-pointed with a weather struck and cut pointing. Describe how this work should be carried out and state what special points must be observed during the operations to ensure a first class job of work.
12. A building has been subjected to serious flooding but the water has now receded. State what particular points in the building should be checked before it is declared fit once again for inhabitation.
13. A wall is found to be very wet due to a defective down pipe which has been leaking for a considerable length of time and the joints have now become soft and pappy. Describe the procedure that should be adopted to rectify this defect.
14. The interior of a building is found to have serious cracks in the solid ground floor. There are also cracks in the partitions, the door frames are out of square and it is difficult to open or close the doors. The external walls appear to be quite sound and the damage seems to be confined to the inside of the building. You are asked to investigate this problem.
    State what you think are likely causes of these defects and what action should be taken to rectify these.
15. An old stonewalled cottage has been rendered with an external decorative wall finish. The interior faces of the external walls of the building which, before the external wall treatment, were quite dry, are now showing signs of dampness. State the likely causes of this dampness and describe the remedial action that may have to be adopted.
16. The inside leaf of an external cavity wall is found to have damp patches at various places on its surface. List the possible causes of this damp penetration and what action may have to be taken to rectify the faults.
17. A niche constructed of gauged brickwork has been badly damaged and it is required to be replaced. Describe the method of cutting and resetting this niche in gauged work.
18. A flint wall having brick dressings to quoins and reveals has become badly damaged and it is necessary to carry out renovation work to restore the wall to its original state. Describe how this work would be executed and how the new work would be tied into the existing walling.
19. A building has been seriously damaged by fire. State what defects are likely to be seen in the walling of the structure as a result of the fire and what action may be necessary to stabilise the walling for future use.
20. A building is found to have serious cracking running roughly diagonally across the face of the building. Some large trees are situated near to the building. State the likely causes of the cracking and how you would carry out an investigation and what course of action you would recommend to rectify the damage.

A Selection of Books on Building Restoration

DECORATIVE PLASTERWORK REPAIR AND RESTORATION   Stagg and Masters

The only book on this important aspect of building restoration, now in its second edition. The text deals with all aspects of the work, both internal and external and is fully illustrated throughout. The text is written for both craft and professional use.

A4 Paperback   ISBN 0-948083-06-9

MASONRY CONSERVATION AND RESTORATION   A.S. Ireson

A new book by the co-author, with Alec Clifton-Taylor, of the best-selling *English Stone Building*. A complete, authorative and fully illustrated text by a master-builder and stonemason and an established authority on the subject. The text includes advice on the selection and preparation of stone for restoration work, the assessment and specification of remedial works and practical methods for the organisation and carrying out of such work. The book includes a glossary of technical masonry terms and a bibliography. Introduction by Sir Bernard Fielden, FRIBA.

A4 Paperback   ISBN 0-948083-04-2

PRACTICAL MASONRY   Wm. R. Purchase

A new edition of an established text by a master-mason, long out of print. The text describes, with the aid of full page plates, the skills required to set out and prepare stone for a wide variety of masonry situations. A complementary volume to *Masonry Conservation and Restoration*. Introduction by John Fidler, RIBA, of *English Heritage*.

A5 Paperback   ISBN 0-948083-07-7

THE STAIRBUILDER AND HANDRAILER   Robert Riddell

A fascimile reprint of the best known text dealing with the design, setting out and construction of a wide variety of timber staircases and their handrails. This edition includes twenty seven full page plates which fully complement the text. The edition is limited to 500 copies of which 475 individually numbered copies are for sale.

A4 Paperback   ISBN 0-948083-08-5

JOINERY REPAIR AND RESTORATION   H. Munn

A new edition of this book will be published in 1990 greatly expanding the original text to include an evaluation of traditional construction methods applicable to the restoration of joinery and the use of both machine and hand working and the constraints placed by both methods on design and construction. Includes schedules of available materials and components.

A4 Paperback   ISBN 0-948083-10-7

Published and sold by
Attic Books The Folly Rhosgoch Painscastle Builth Wells Powys LD2 3JY